"This stellar collection digs deep into prior scholarship on resource extraction and unearths bold new agendas for future research."

Stuart Kirsch, author of *Mining Capitalism*

"A nuanced, rich account, *The Anthropology of Resource Extraction* brings together leading scholars to make valuable contributions to key theoretical debates, while emphasizing the value of ethnography for understanding the practice of extraction."

Eleanor Fisher, Head of Research, Nordic Africa Institute

"*The Anthropology of Resource Extraction* is certain to become a classic for scholars interested in the entangled social and environmental worlds of humans and non-renewable minerals. Taking a thematic approach to the analysis of our utter dependence upon minerals for contemporary livelihoods, this volume brings together leading scholars in anthropology to review the state of resource extraction today. Topics ranging from corporations, to the environment, to governance, to water, are treated with a fine-grained ethnographic attention to the explosion of interest in extractivism in the 21st century."

Jerry K. Jacka, Associate Professor of Anthropology, University of Colorado Boulder, U.S.A.

THE ANTHROPOLOGY
OF RESOURCE EXTRACTION

This book offers an overview of the key debates in the burgeoning anthropological literature on resource extraction.

Resources play a crucial role in the contemporary economy and society, are required in the production of a vast range of consumer products and are at the core of geopolitical strategies and environmental concerns for the future of humanity. Scholars have widely debated the economic and sociological aspects of resource management in our societies, offering interesting and useful abstractions. However, anthropologists offer different and fresh perspectives – sometimes complementary and at other times alternative to these abstractions – based on field researches conducted in close contact with those actors (individuals as well as groups and institutions) that manipulate, anticipate, fight for, or resist the extractive processes in many creative ways. Thus, while addressing questions such as: "What characterizes the anthropology of resource extraction?", "What topics in the context of resource extraction have anthropologists studied?", and "What approaches and insights have emerged from this?", this book synthesizes and analyses a range of anthropological debates about the ways in which different actors extract, use, manage, and think about resources.

This comprehensive volume will serve as a key reading for scholars and students within the social sciences working on resource extraction and those with an interest in natural resources, environment, capitalism, and globalization. It will also be a useful resource for practitioners within mining and development.

Lorenzo D'Angelo is Assistant Professor in the Department of History Anthropology Religions Art History, Media and Performing Arts, Sapienza University of Rome, Italy.

Robert Jan Pijpers is a Postdoctoral Researcher at the Institute of Social and Cultural Anthropology, University of Hamburg, Germany.

ROUTLEDGE STUDIES OF THE EXTRACTIVE INDUSTRIES AND SUSTAINABLE DEVELOPMENT

For more information about this series, please visit: www.routledge.com/Routledge-Studies-of-the-Extractive-Industries-and-Sustainable-Development/book-series/REISD

THE ANTHROPOLOGY OF RESOURCE EXTRACTION

Edited by
Lorenzo D'Angelo and Robert Jan Pijpers

Routledge
Taylor & Francis Group
LONDON AND NEW YORK

earthscan
from Routledge

Cover image: Getty Images

First published 2022
by Routledge
605 Third Avenue, New York, NY 10158

and by Routledge
4 Park Square, Milton Park, Abingdon, Oxon OX14 4RN

Routledge is an imprint of the Taylor & Francis Group, an informa business

British Library Cataloguing-in-Publication Data
A catalogue record for this book is available from the British Library

Library of Congress Cataloging-in-Publication Data
Names: D'Angelo, Lorenzo, editor. | Pijpers, Robert Jan, editor.
Title: The anthropology of resource extraction / edited by Lorenzo D'Angelo
and Robert Jan Pijpers.
Description: Abingdon, Oxon ; New York, NY : Routledge, 2022. |
Series: Routledge studies of the extractive industries | Includes bibliographical
references and index.
Identifiers: LCCN 2021030892 (print) | LCCN 2021030893 (ebook) |
ISBN 9780367862596 (hardback) | ISBN 9780367687533 (paperback) |
ISBN 9781003018018 (ebook)
Subjects: LCSH: Mineral industries–Social aspects. | Mines and mineral
resources--Social aspects. | Anthropology.
Classification: LCC HD9506.A2 A58 2022 (print) | LCC HD9506.A2 (ebook) |
DDC 338.2--dc23
LC record available at https://lccn.loc.gov/2021030892
LC ebook record available at https://lccn.loc.gov/2021030893

ISBN: 978-0-367-86259-6 (hbk)
ISBN: 978-0-367-68753-3 (pbk)
ISBN: 978-1-003-01801-8 (ebk)

DOI: 10.4324/9781003018018

Typeset in Bembo
by Taylor & Francis Books

CONTENTS

CONTRIBUTORS

Nicholas Bainton is an Associate Professor at the University of Queensland. He has written widely on the social and political effects of extractive capitalism in Melanesia and is the author of *The Lihir Destiny: Cultural Responses to Mining in Melanesia* (2010), and editor of *Unequal Lives: Gender, Race and Class in the Western Pacific* (2021), and with Emilia Skrzypek *The Absent Presence of the State in Large-Scale Resource Extraction Projects* (2021). ORCID: 0000-0001-7571-3679

Jeroen Cuvelier is a Senior postdoctoral research fellow of the Fund for Scientific Research-Flanders at the Department of Conflict and Development Studies of Ghent University. Between 2017 and 2021, he also served as a visiting professor at the Institute for Anthropological Research in Africa (IARA) of KU Leuven, where he taught courses in economic anthropology and the anthropology of development. His current research deals with the politics of time in environments where large-scale and small-scale forms of mining co-occur. ORCID: 0000-0001-5772-1691

Lorenzo D'Angelo is Assistant Professor at the Sapienza University of Rome, Italy and a Research Fellow at the University of Hamburg, Germany. He has carried out fieldwork in Sierra Leone, Tanzania, and Uganda. He is the co-founder of the EASA Network Anthropology of Mining, co-editor of the special journal issue *Mining Temporalities* (2018) and the author of the book *Diamanti. Pratiche e stereotipi dell'estrazione mineraria in Sierra Leone* (2019). ORCID: 0000-0002-8731-2714

Elizabeth Ferry is Professor of Anthropology at Brandeis University, with interests in value, materiality, mining, and finance, and with fieldwork emphases in Mexico, Colombia, the USA and the UK. Among other works, she is the author of *Not Ours Alone: Patrimony, Value and Collectivity in Contemporary Mexico* (2005), co-author of *La Batea* (2017) and co-editor of *The Anthropology of Precious Minerals*

(2020), and of articles in *Cultural Anthropology, Economy and Society, The Journal of Material Culture*, and other journals.

Sara Geenen is Assistant Professor at the Institute of Development Policy (IOB), University of Antwerp, Belgium. She is project leader of the Centre d'Expertise en Gestion Minière (CEGEMI) at the Université Catholique de Bukavu, DRC. Her research interests lie in the global and local development dimensions of extractivist projects, addressing questions about more socially responsible and inclusive forms of globalization. ORCID: 0000-0002-0933-3974

Cristiano Lanzano is Senior Researcher at the Nordic Africa Institute in Uppsala (Sweden). He has conducted research on environmental conservation, natural resource management and small-scale mining in West Africa. He focuses on informal governance, technological and social change, mobilities, and sustainability issues in the artisanal gold mining sector. He has published on these topics in journals such as *Politique Africaine, Autrepart, The Extractive Industries and Society*, and the *Journal of Agrarian Change*. ORCID: 0000-0003-0481-5250

Fabiana Li is an Associate Professor of anthropology at the University of Manitoba, Canada. Her research focuses on the politics of knowledge relating to the environment, agriculture, and social movements in Latin America. She is the author of *Unearthing Conflict: Corporate Mining, Activism, and Expertise in Peru* (2015) as well as articles on corporate social responsibility, the anthropology of water, and resource conflicts. ORCID: 0000-0001-7722-3067

Sabine Luning is Associate Professor at the Institute of Cultural Anthropology and Development Sociology, Leiden University, The Netherlands. Her research interests include economic anthropology, infrastructure, and the nexus between resource extraction and water management. She has carried out research on large-scale and small-scale gold mining, mainly in Burkina Faso and Ghana, but also in Suriname, French Guyana, and Canada. She is currently one of the leads in the Gold Matters project and the Leiden-Delft-Erasmus PortCityFutures Program. ORCID: 0000-0002-9951-2780

Mark Nuttall is Professor and Henry Marshall Tory Chair of Anthropology at the University of Alberta. He has done fieldwork in Greenland, Canada, Alaska, Finland, Scotland, and Wales and his research interests include climate change, energy and resource development, conservation, and political geology. He is author of *Climate, Society and Subsurface Politics in Greenland: Under the Great Ice* (Routledge 2017), co-author of *The Scramble for the Poles* (Polity, 2016), and co-editor of *The Routledge Handbook of the Polar Regions* (2018). ORCID: 0000-0002-1620-3959

Robert Jan Pijpers is a Postdoctoral Researcher at the University of Hamburg. His work interrogates processes of social, economic, and environmental change, as well as (re)configurations of power relations in the context of resource extraction in West Africa. He has co-founded and convened two international networks of scholars working on extraction, has published articles in *Africa, Anthropology Today*, and *History and Anthropology*, among other journals, and is the co-editor of *Mining Encounters: Extractive Industries in an Overheated World* (2018). ORCID: 0000-0002-2031-1178

Dinah Rajak is a Reader in Anthropology and International Development at the University of Sussex. She is the author of *In Good Company: An Anatomy of Corporate Social Responsibility* (2011), co-editor of *The Anthropology of Corporate Social Responsibility* (2016), and co-founder of the Centre for New Economies of Development.

Benjamin Rubbers is Professor of Social Anthropology at the Université de Liège, Belgium. He is the author of several books and articles on the Congolese copperbelt, where he has done research since 1999. His research interests focus on work, mining, capitalism, and the local state in Africa. His last book publication is a co-edited volume with Alessandro Jedlowski: *Regimes of Responsibility in Africa. Genealogies, Rationalities and Conflicts* (2019). ORCID: 0000-0002-2399-7937

Elana Shever explores the nexus of science, industry, and capitalism in her research. She is the author of *Resources for Reform: Oil and Neoliberalism in Argentina* (2012), and is currently writing an ethnography of palaeontology in the United States. She is Associate Professor of Anthropology at Colgate University, and has been a Scholar-in-Residence at the University of Minnesota's Institute for Advanced Study, and a Fellow at the University of Rochester Humanities Center and Brown's Watson Institute for International Studies. ORCID: 0000-0001-9112-5914

Emilia E. Skrzypek is a Senior Research Fellow at the University of St Andrews. Her work to date has largely focused on Papua New Guinea, where she investigates issues related to broadly conceived resource relations and interdependencies. She is the author of *Revealing the Invisible Mine: Social Complexities of an Undeveloped Mining Project* (2020), and editor with Nicholas Bainton of *The Absent Presence of the State in Large-Scale Resource Extraction Projects* (2021). ORCID: 0000-0001-6357-0180

Teresa A. Velásquez is an Associate Professor of anthropology at California State University, San Bernardino. Her ongoing research examines forms of protest and contestation in Ecuadorian mining conflicts. She is currently finishing her book manuscript titled *The Water is Ours: Mining and the New Politics of Indigeneity in Ecuador*. Her publications appear in *Latin American and Latinx Visual Culture, Latin American Perspectives, Latin American and Caribbean Ethnic Studies*, and *Resources Policy*, in addition to contributions in edited volumes. ORCID: 0000-0002-2579-4383

Boris Verbrugge is a Senior Researcher at the HIVA research institute (KU Leuven). He is involved in a range of research projects on business and human rights, and has a particular interest in understanding how initiatives in the field of "sustainable" and "transparent" supply chains differentially affect actors operating along global value chains. ORCID: 0000-0002-4773-551X

PREFACE

Resource extraction is growing and so is its anthropological interrogation. In recent decades, anthropological contributions to understanding the multiple social, political, economic, historical, cosmological, and environmental dynamics associated with resource extraction have expanded significantly. These studies offer crucial insights into the politics, processes, and effects that are imbricated in an activity that confronts all of us on a daily basis, whether consciously or unconsciously, directly or indirectly. This volume brings together a large range of research and serves as a reference point for scholars, professionals, and students alike. Our main aim, however, is to consolidate this growing body of work into a coherent field of anthropological inquiry and to sketch the contours of what we understand as the "The Anthropology of Resource Extraction". In this sense, the volume follows on from earlier initiatives we have taken in this direction, including the establishment of the EASA Anthropology of Mining Network in 2014. More and other kinds of initiative along similar lines have been (and will hopefully continue to be) taken, and we expect that this volume will offer a solid contribution to this common effort.

Of course, such initiatives, whether the establishment and running of networks or the publication of editorial collections like this one, are always the result of multiple collaborations. These include the formal and informal conversations and other engagements with colleagues during conferences, the processes of working together towards joint outputs, departmental seminars and "coffee-machine talks" and, of course, the many kinds of interaction that take place within the institutes and the research projects in which we have been embedded, at the time of writing this publication as well as in the past. While this volume was being prepared, we were both part of the Gold Matters project – a transdisciplinary research initiative which considers whether and how a transformative approach towards sustainability can develop in artisanal and small-scale gold mining (ASGM). Some of our Gold

Matters colleagues are represented in this volume through their contribution of a chapter, but all have been part of our academic engagements over recent years and, as such, also of our thought processes. We are very grateful for their stimulating input.

The book itself is also the result of multifarious collaborations and, accordingly, we have greatly appreciated the many dialogues with the various contributing authors. To some extent, producing this book has turned into a sort of laboratory of ideas triggered by conversations following the reading of each other's chapters and providing feedback. Indeed, not only have the authors in this volume engaged with the large body of published work of colleagues with respect to their own chapters, they have also provided critical comments on the other chapters in this book. Such investments of time can, of course, not be taken for granted; not only are they invaluable, they are also always conducted on top of our other work.

This has been all the more pertinent during the demanding times in which this volume materialized: that of the COVID-19 pandemic, which started to affect daily affairs globally from early 2020, is still disruptive as we write this preface and is likely to continue to be so for some time. For us, as academics, it has sometimes meant conducting our teaching in digital format, attending virtual meetings and conferences, and, in some cases, carrying out our research online. It has also been a time of home schooling, travel restrictions, providing care for our families, and mourning losses, sometimes at a distance. A turbulent period, which makes us all the more appreciative of the importance of a book that explores the complexities, paradoxes, contestations, and ambivalences of this omnipresent yet sometimes taken for granted element of life: resource extraction. We hope that this book also represents a contribution to the understanding of the times in which we live and the ways in which different people are imbricated, perhaps caught up, in this key aspect of contemporary, globalized society.

Finally, the work we, as anthropologists, do, would be impossible without the dedicated time and energy of those people who so generously share their experiences with us, and offer insights into their lives, interests, hopes, challenges, and concerns. Ultimately, it is thanks to these collaborations, the access granted (or denied) and the privileges of being allowed to gain insight into the lives and life-worlds of a wide variety of people, that we are able to interrogate how resource extraction makes the world go round. This book is dedicated to all of them.

Lorenzo D'Angelo & Robert Jan Pijpers
Milan, Italy & Hamburg, Germany

1

THE ANTHROPOLOGY OF RESOURCE EXTRACTION

An Introduction

Lorenzo D'Angelo and Robert Jan Pijpers

1 Introduction

The extraction and use of resources has long been a defining factor for society, as evidenced by the links that have been established between natural resources and certain phases of human history. Expressions such as "stone age", "bronze age", or "iron age" underline the determining role of particular technologies based on specific materials in conditioning the economic and political organization of past societies (see Boivin and Owoc 2004; Knapp et al. 2002). Similarly, yet in a more contemporary context, scholars have coined terms such as "petrocapitalism" (Huber 2017) and "carbon democracy" (Mitchell 2011) to understand how dependence on fossil fuels affects our global political and economic organization. Jacka (2018), on the other hand, proposes the more all-encompassing term "mineral age" to describe the current historical moment which stands out for its unprecedented levels of production and consumption of mineral resources (see also Arsel et al. 2016) and in which "our entire livelihoods are utterly dependent on minerals" (Jacka 2018: 62).

This utter dependence on minerals that defines our current world and time is indirectly illustrated by a recent study published in *Nature* (Elhacham et al. 2020). For the first time in history, this study certifies, the mass of human-made things exceeds all living biomass on Earth, comprising the mind-boggling volume of about one teraton (10^{18}gr). While this study draws attention to the force and scale which human beings now bring to shaping the environment and Earth systems, a situation encapsulated by many scholars in the concept of the Anthropocene (e.g. Steffen et al. 2007), it does not consider the processes of extraction that underpin the creation of these human artefacts. After all, these human-made things – bridges, roads, houses, stapling machines, cars, planes, cell phones, pens, dams, electricity poles, traffic lights, sports medals, heaters, railway tracks, and so forth – all require

DOI: 10.4324/9781003018018-1

the extraction of mineral resources ranging from iron ore, sand, copper, tin, tantalum, lithium, and gold to oil, gas, coal, and uranium.

The processes of extraction that are needed to recover these and many other mineral resources are as diverse as they are complex. They materialize in (and connect) different social, political, historical, and economic terrains, use different technologies, assume different scales (ranging from the artisanal to the industrial) and involve a wide variety of actors. These actors are integrated into different networks, take up different positions and have different ideas, perceptions, and understandings, as well as different interests, dependencies, and powers. Crucially, they all exercise a certain degree of influence on shaping the processes of resource extraction, albeit on highly unequal terms. It is exactly these processes, and how they are entangled in – that is, both shape and are shaped by – the social relationships and lifeworlds of hundreds of millions of people globally, that are at the centre of this volume on the anthropology of resource extraction.

2 Understanding a global affair

Unmistakeably, resource extraction is more than ever a global, planetary affair (cf. Arboleda 2020; Labban 2014), subsuming increasingly large volumes of earth, land surface, and numbers of people. Resonating with processes of acceleration and exponential growth that can be observed in contemporary society more broadly (Eriksen 2016), the global annual volume of extracted materials has tripled in the last 50 years, growing from 27 billion tons in 1970 to 92 billion tons in 2017 (Oberle et al. 2019) – the equivalent of more than 9.1 million Eiffel Towers. In the last two decades or so, this growth has been particularly rapid due to the most recent commodity super-cycle, that is, a period marked by unusually high demand for and price increases in primary commodities, typically, but not exclusively, related to processes of industrialization (e.g. Burton et al. 2021; Ericsson and Löf 2019). Indeed, the most recent of such cycles began at the turn of the millennium and was largely driven by industrialization processes in the so-called BRIC countries (Brazil, Russia, India, China), particularly in China. While this cycle has just ended, the next one is already knocking on planet Earth's door, and is this time likely to be linked to global "green" or "clean" energy and technology agendas that depend on the extraction of ever increasing quantities of minerals such as nickel, lithium, copper, and cobalt (IEA 2021). Not surprisingly, along with population growth and consumption patterns, the OECD anticipates that over the next 40 years the global annual amounts of extracted materials will double (OECD 2019).

All these materials require different types of operation to be extracted. However, a common aspect of all extractive activities – whether carried out by artisanal miners using spades and buckets or large-scale companies employing automated excavators – is "the need to move earth" (Bridge 2004: 209). It follows that the large volumes of extracted resources mentioned above require the movement of an even greater amount of earth. Illustratively, it is estimated that to obtain an ounce of gold, 30 metric tons of rock need to be processed, while some of the largest mines in the world move up to half a million metric tons of earth a day in order to obtain these 30 tons (Perlez and Johnson 2010). All this movement of earth is estimated to potentially have an impact on 50 million km^2 of our planet's surface (Sonter et al. 2020; see also Maus et al. 2020) – which is more than the surface of Africa and Latin America combined. Among the territories concerned are those that have an important ecological role (Oberle et al. 2019) and, according to the United Nations International Resource Panel, 90% of biodiversity loss and global water stress is caused by the extraction and processing of natural resources, while the metal and mineral industries contribute to about 50% of the emissions of greenhouse gases, that is, those gases that are most worrying because of their role in atmospheric and climate changes (ibid.).

This large-scale significance of resource extraction also features in other domains. In many (mostly low and middle-level income) countries, for example, extraction takes a more-than prominent position in national economies. In Suriname, almost 20% of the country's total GDP derives from mineral rents, while in the Democratic Republic of Congo, Mongolia, and Bolivia, about 91%, 85%, and 43% of total export values, respectively, are extraction-related (ICMM 2020). Moreover, in terms of livelihood opportunities and employment, it is estimated that artisanal and small-scale mining (ASM) offers a direct livelihood opportunity to approximately 40 million people globally, while the number of those dependent on ASM is estimated at 150 million (IIED 2017). Moving from the artisanal and small-scale to the industrial scale, and from global to national-level figures, shows that in 2017, for example, the minerals sector employed 634,000 individuals in Canada,[1] 240,000 in Australia[2] and 457,698 in South Africa, where another 4.5 million people are being considered so-called dependents.[3] Yet, in contrast with this large number of people to whom extraction offers an opportunity, many also face severe challenges and risks due to the extraction of resources within or near their living environments. In this regard, the number of conflicts reported in the Environmental Justice Atlas (Temper et al. 2015), although far from complete, as well as the estimation that coal mining in India alone caused the displacement of more than 2.55 million people between 1950 and 1990 (Downing 2002), are as equally telling as they are striking.

Taking all of the above into account, it is clear that "the power of the mineral world is difficult to ignore" (Boivin 2004: 1), and understanding the wide variety of dynamics that are connected to extraction seems to be more pertinent than ever before. The question is how to give shape to this attention and grasp the full complexities and effects of resource extraction. After all, as much as these flows, volumes, and numbers reveal that resource extraction is a factor that concerns humanity as a whole – and one marked by high degrees of inequality, contestation, and ambivalence, to say the least – they can only offer abstract, incomplete, and homogenizing representations of what are complex and variegated processes. As such, these flows, volumes, and numbers reflect both a disconnection of extractive environments from larger production systems (Bonneuil and Fressoz 2016; see also

Bunker 1984; Ciccantel and Smith 2009) and a "depersonalisation that has dominated accounts of the political economy of resource extraction" (Gilberthorpe and Rajak 2016: 8). Consequently, many of the dynamics that inform *how* resource extraction is entangled in social, political, economic, and physical environments and *how* a wide variety of actors attempt to shape this entanglement are left unaddressed or even obscured.

One may wonder, for instance, exactly what the environments in which extraction materializes look like. What is the range of values and moralities attributed to them and the mineral resources that they contain? What livelihood activities do these environments enable or foreclose? What are the histories of these environments, and what role does extraction play in these histories? Who are the people inhabiting and giving shape to these environments? What are their aspirations, dreams, challenges, and fears? And how does extraction, which generates both opportunities and extensive forms of disruption, affect their lives in all their different facets? These questions point to the social worlds in which extraction is situated – and which are, to varying degrees, shaped by extraction – as well as the processes through which extraction materializes and comes to exist in these social worlds. After all, extractive practices are dependent on a range of processes whereby transnational corporations or individual artisanal miners attempt to seek *and* maintain access to specific terrains and the minerals they contain. How are such processes shaped, negotiated, and renegotiated? On what (multi-layered) governance structures are they dependent? Which actors are involved? How do they relate to each other and what are their different interests, concerns, desires, and powers?

Asking such questions and addressing such issues are key to understanding how, to quote Eriksen (2016: 8), "global processes interact with local lives in ways that are both similar and different across the planet". In the case of extraction, it untangles and enables understanding of how and why extraction assumes the various shapes and produces the various effects that it does. Without ignoring the view from above, interrogating this requires a view "from below" (Mathews et al. 2012) and possibly from within, one that takes the environments and relationships (including those with non-human actors) of the people involved in global extractive capitalism as its starting point. The anthropology of resource extraction does exactly this.[4] It embeds resource extraction in the particular sets of histories and social, political, and economic relations of specific localities, and places its social relations "front and centre" (Gilberthorpe and Rajak 2016: 8), thereby staying attuned to the wide variety of extractive "encounters" and concurrent negotiation processes (Pijpers and Eriksen 2018) through which social relations, as well as the forms and effects of extraction, are (re)shaped. In doing so, the anthropology of resource extraction serves "to document and critically theorize the interplay of capital, corporations, states, and place-based livelihoods, and the impact it [extraction] has on the environment" (Jacka 2018: 63) and to investigate "how the economies of extraction create new domains for the exercise of power, and new struggles over authority, at the micro-level as much as the macro" (Gilberthorpe and Rajak 2016: 8). In the broadest terms, we would add to the above, the anthropology of resource extraction interrogates the diverse range of shapes that resource extraction takes and *how* it is entangled in and affects specific environments, lifeworlds, and relationships. Ultimately, it aims to contribute to understanding the manifold dimensions of life in the contemporary world by factoring in the role played by resource extraction.

3 Approaching extraction

Before we proceed, it is important to clarify what we mean by "resource extraction". In its most ordinary understanding, extraction refers to "the process of removing something, especially by force", "the process of removing or taking out something", or, the most relevant in our case, "the process of removing a substance from the ground or from another substance".[5] Under certain circumstances, these substances are extracted for their ascribed value and can be considered resources. The word "resource" draws its meanings from the French *ressource*, the Old French *ressourdre, resourdre* (to spring forth or up again)[6] which, in turn, come from the Latin *resurgo* (to rise again, spring up anew).[7] The extractive industries considered in this volume, however, deal with non-renewable resources, that is, substances that cannot rise again or spring up anew. It is worth stressing that the non-renewability of resources does not imply that they are fixed entities. What a resource is or, better, how a resource *becomes* through processes of valorization (see also discussions by both Ferry and Luning this volume), depends on variations of historical and geographical circumstances as well as on specific technical possibilities and the cultural, economic, and political needs of a given society (Bridge 2009: 1220).[8] Thus, to rephrase Gavin Bridge, resource extraction is a process of material transformation through which certain substances that are considered valuable are separated from other substances and environments to be converted into a vast array of commodities or artefacts (ibid.: 1221).

Recently some scholars have extended this understanding of extraction in ways that foreground the logics, ideologies, and inequalities that underpin extraction (and more broadly, capitalism), thereby directing attention to the idea of "extractivism" (Gudynas 2010; see, among others, Acosta 2013, 2014; Svampa 2019). In the political-economic and political-ecological literature on South America, extractivism generally refers to a system of accumulation and processes that transfer natural resources from the peripheries to the centres of the world economy and that can be traced back to imperialism and colonialism yet persist in the present (Gudynas 2010; Jacka 2018; Parks 2021; Szeman and Wenzel 2021). In the wake of this literature, scholars have also used the terms neo-extractivism or new extractivism to account for a more specific development observed in some South American countries since the early 2000s (see Gudynas 2010). Here, the narrative that identified the extractive industries' rationale as one furthering predatory capitalism, private corporations' interests, and a neoliberal economy no longer sufficed, given that some national governments ruled by left-wing parties reoriented policies in ways that rejected neoliberal models, encouraged nationalization processes, and focused on using the profits obtained from extractive activities to finance social policies aimed at, for example, reducing poverty. The terms neo-extractivism and new extractivism take on this reoriented ideological connotation in which "the extraction of raw materials is socio-politically justified by the necessity to struggle against poverty and social inequalities" (Brand et al. 2016: 130). Despite some differences, these neo-extractivist approaches are largely analysed and critiqued along the same lines as older forms of extractivism as they are equally based on the appropriation of nature, dependent on international involvement, and accompanied by significant environmental and social impacts (Gudynas 2010: 1).

As transpires from the above, the concepts of extractivism and neo- or new extractivism stress extraction as an "ideological mindset" (Jalbert et al. 2017b), approaching the *act* of resource extraction as a policy, ideology, or logic of extractivism (Szeman and Wenzel 2021: 508). This approach or understanding allows scholars to foreground and critically interrogate the ideologies, relationships, and structures of inequality, as well as the destruction and exploitation of nature that underpin the extraction of resources.

The term "extractivism" (as well as "extraction"), however, is also increasingly applied to domains beyond the extraction of minerals, including agriculture, finance, logistics, fishing, and tourism (e.g. Acosta 2017; Gómez-Barris 2017; Mezzadra and Neilson 2017; Mollona 2019). In fact, its usage is becoming so widespread that Szeman and Wenzel (2021: 505) critically point out that extractivism "has quickly become the name for every process and practice through which value is generated for capitalism". This more generic or abstract usage of the notions of extractivism and extraction uncovers important relations of inequality associated with global capitalism more broadly. Yet, without going into detail and notwithstanding potentially fruitful points of comparison, there are of course important differences between the experiences and the broader social, economic, material, and political worlds of a miner *extracting* a mineral resource from the underground, a financial analyst or a digital miner using complex algorithms to *extract* data, a member of an indigenous community whose heritage is *extracted* by foreign tourists, a literature researcher *extracting* information from an archive, or an employer that *extracts* value from workers' labour. Considering these differences, some scholars have pointed out the potential risks of "metaphorical inflation" and "loss of analytical precision" when applying a more metaphorical, expanded, or abstract usage of the notions of extractivism and extraction (ibid.). Indeed, if searching for a generalizable principle in practices so *materially* different as "extracting" data and extracting minerals results in having an empty signifier that is applied almost everywhere (that is, "extractivism" or "extraction" in an expanded sense that encompasses all value-creation processes under capitalism), how do we ensure that we fully understand, anthropologically, the specificity of each of these practices and their associated materialities, imaginaries, and social relations?

Taking the above considerations into account, and to be clear about the contours of this volume, our departure point is a more "conventional" or literal approach to resource extraction: the removal of non-renewable resources – metallic and non-metallic minerals, oil, gas, and coal included – and the various social, political, economic, legal, and technical processes involved. This approach implies that while we "limit" ourselves to the *acts* or *practices* of resource extraction and the multifarious dynamics underpinning them, we also open up to a wide range of dynamics and effects that can be revealed through an ethnographic analysis of the lived and embedded experiences of social actors involved in resource extraction who all bring their specific individual and collective hopes, desires, wishes, plans, concerns, and worries. Consequently, while simultaneously being a way of narrowing down to a specific sense of the word "extraction" as well as opening up a diverse range of dynamics and specificities, we believe that our approach offers a framework that effectively and productively delineates resource extraction as an object/field of anthropological study.

4 A brief reflection on an exploding field

Given its prominence throughout history, one might expect that resource extraction would have always played an important role in anthropological reflections. Yet anthropological interest in the extractive industries emerged relatively late, as Godoy stressed in the first annual review of the study of mining in anthropology in 1985. Since then, however, a significant body of work has been published, including several literature reviews (Ballard and Banks 2003; Cuvelier et al. 2014; Gilberthorpe and Rajak 2016; Godoy 1985; Jacka 2018; Rogers 2015) and encyclopaedia entries (Bainton 2020; Golub 2019; Smith and Kirsch 2018), a number of *ethnography-based* "anthropologies of" – for example, precious minerals (Ferry et al. 2019), oil (Behrends et al. 2011), energy (Smith and High 2017), and corporate social responsibility (Dolan and Rajak 2016) – as well as numerous insightful monographs, articles, and edited volumes. In what follows we focus our attention on how this large body of anthropological scholarship has been examined in four so-called "annual reviews". We then briefly consider some of the most recent edited works published in the last two decades to show the main directions taken by the academic debate.[9]

Mining: anthropological perspectives (1985)

In the mid-1980s, Ricardo Godoy (1985) attributed the then recent anthropological interest in extraction to, on the one hand, the environmental and energy crisis of those years and, on the other, the theoretical debates on the limits of resources and industrial development that matured from the 1960s onwards. His annual review examined those anthropological studies that considered mineral economics; the demographic, social, and political characteristics of mining communities; and the rituals, beliefs, and ideology associated with mining. Godoy argued that anthropologists have largely focused on the second two domains, which he labels mining's derivative sociopolitical and ideological superstructures, and in particular on the "the causes, consequences, and meaning of migration to the mines" (ibid.: 211). Despite an emerging interest in mining, Godoy concludes that "systematic studies have yet to arrive" and "one is not likely to find ideas, much less a coherent system of ideas, in anthropological studies of mining" (ibid.: 199). However, notwithstanding this rather critical assessment of a yet-to-emerge anthropology of mining or anthropology of resource extraction, his review discussed a number of works that would prove to have lasting influence on the anthropology of resource extraction. Worth mentioning here is the work of June Nash and Michael Taussig on Bolivian tin miners and the effects of their integration into a global capitalist economy, which draws attention to processes of cultural transformation (Nash [1979] 1993) and experiences of alienation (Taussig 1980). Moreover, across the Atlantic, scholars associated with the Rhodes Livingstone Institute in colonial Rhodesia (many of whom would become associated with the Manchester School), examined processes of social change spurred by industrial development in the Copperbelt, thereby particularly foregrounding transitions from tribal/rural to modern/urban societies (Wilson 1941; Mitchell 1956; Epstein 1958; Gluckman 1961) (for analytical reflections see also Falk-Moore 1994: 50–51).

Pointing at the gaps and shortcomings within studies of mining (mostly in anthropology, but also beyond) Godoy invited scholars to take an approach that examines "both the geological and economic infrastructure of the firm/industry as well as their secondary socio-political and ideological dimensions" (1985: 211). Moreover, "given the anthropologists' penchant for, and strengths with, small-scale social systems", he argues, "the major and first impact of anthropological contributions is likely to be made in the small-scale mining sector" (ibid.). Many anthropologists have indeed focused on this sector, although this focus has certainly not been an exclusive one, as we will see in the next sections.

Resource wars (2003)

Almost two decades after the publication of Godoy's annual review of the anthropology of mining, Chris Ballard and Glenn Banks (2003) conducted a second one, focusing attention on anthropological studies that examine the link between mining and conflict (exemplarily captured by its title "Resource Wars"). They note that in the nearly 20 years since the publication of the first systematic anthropological review of mining studies, many things had changed. The minerals boom of the 1980s, the same one observed by Godoy, had had an enormous effect on the global mining industry which appears to have expanded like never before. Mining had begun to affect ever enlarging, so-called "greenfield" territories (simply put, new territories) and communities, in particular indigenous ones, which, traditionally, have been the privileged research locus for many ethnographers (although see also Bridge 2004). Subsequently, these communities quickly assumed a "pivotal position in the politics and analyses of the wider global mining community"

(Ballard and Banks 2003: 288), thereby widening a previously binary contest between states and corporations. Congruently, the idea of such a mining community was extended to incorporate the growing variety of actors that coalesce around projects, while the internal complexity of specific actors, such as corporations, states, and communities, also became increasingly foregrounded. In fact, one of the two key aims of Ballard and Banks's review was to question "the often-monolithic characterizations of state, corporate, and community forms of agency" (ibid.: 287).

Within this sphere of an accelerated mining industry, the incorporation of new territories and communities, and the increasingly complex shapes that mining communities could take, a key point of departure for Ballard and Banks is that "mining is no ethnographic playground" (ibid.: 289). On the contrary, as also transpires throughout this volume, "relationships between different actors within the broader mining community have often been characterized by conflict, ranging from ideological opposition and dispute to armed conflict and the extensive loss of lives, livelihoods, and environments" (ibid.: 289).

This contested nature prompts Ballard and Banks to chart the debate over the politics and ethics of engagement by anthropologists involved in mining, whether as consultants, civil society advocates, or researchers. Moreover, as mining environments are not only contested, but also increasingly complex, Ballard and Banks conclude their review by relaunching Godoy's plea for "an 'integrative' approach to the anthropology study of mining" (ibid.: 307). Their proposal envisions a dialogue between anthropology and other disciplines in which anthropology relies on multi-sited ethnography (Marcus 1995) as one of the most promising approaches for understanding contemporary extractive industries, given that mining activities typically span multiple sites which are connected at different spatio-temporal scales and levels (among other considerations). Furthermore, due to the complexities of the extractive industries and the inevitable limitations of individual research, they wish for "flexible coalitions or alliances of often-unlikely partners" (Ballard and Banks 2003: 307) between the academic and the non-academic worlds, including NGOs, industry think-tanks, and community activists.

Social and environmental impacts in the mineral age (2018)

Jerry Jacka's review (2018) sits on a temporal continuum with the two previous anthropological reviews (Ballard and Banks 2003; Godoy 1985), examining, above all, the literature that emerged in the 15 years immediately preceding its own publication. In this period of time, extractive activities have continued to grow, but two trends can be seen as distinctive for the current mode.

First, there is the idea of extractivism as a mode of capitalist accumulation based on the extraction of natural resources that is at least 500 years old (see also discussion in Section 3 of this introduction); however, what "makes extractivism new and imperative is that it is the primary means of socioeconomic development that developing countries themselves pursue, even in left-leaning nations in Latin America" (Jacka 2018: 62). Second, today's resource extraction is conditioned by international financial institutions, mining sector reforms and the liberalization of the mining sector through privatization and deregulation (although some South American countries have countered this trend), and corporations that promise forms of moral and sustainable extraction – a promise that some scholars have classified as an oxymoron (Kirsch 2010). Considering these trends, and the increasing number of lands and people that extraction incorporates, Jacka argues that "ethnographic research on mining serves to document and critically theorize the interplay of capital, corporations, states, and place-based livelihoods, and the impact it has on the environment" (2018: 63).

To illustrate how anthropologists have taken up this task, Jacka centralizes four interrelated domains. First are the livelihood transitions that occur in the context of the expansion of extraction, a focus which, among others things, leads him to engage with questions of livelihood diversification, the effects of sector reforms, and deagrarianization in relation to artisanal and small-scale mining (ASM), a domain that, as Godoy predicted, has indeed become of increasing importance to anthropologists and other scholars alike (Bryceson et al. 2013; Buss and Rutherford 2020; Cremers et al. 2013; Hilson 2002; Hilson and McQuilken 2014; Lahiri-Dutt 2018; Werthmann and Grätz 2012), although certainly not the only one. Second, Jacka considers the transformations in corporate practices that can be observed in recent decades, largely relating this to corporate social responsibility (CSR), and the emphasis placed by anthropologists studying it on the "need to understand how CSR is performed or enacted and ultimately recognized" (2018: 65). This is followed by consideration of the environmental impacts of extraction, incorporating the effects on land, water, and air, and stressing that anthropologists have most studied these impacts "through exploring the changing human–environmental relations brought by mining development" (ibid.: 67). Finally, Jacka discusses a fourth domain where anthropology has made significant contributions: the social conflicts (and, in their wake, also the numerous forms of resistance) that are an omnipresent element of mining environments globally and, crucially, the multi-directional processes through which these conflicts take shape.

Oil and anthropology – temporality and materiality (2015)

At the core of Douglas Rogers's (2015) review, which is focused on oil, lie concerns similar to Jacka's, that is ethnography's task of critically interrogating the interplay of capital, corporations, states, and place-based livelihoods, and the impact that extraction, in this case of oil, has on the environment. Given the economic and geopolitical role that oil has played for more than a century, such a focus seems to be more than justified. As already observed by Boyer (2014), anthropologists have shown interest in this resource and, more generally, in energy, in at least three different historical periods: the beginning of the 1940s, the beginning of the 1970s, and, finally, the current period, which has seen an exponential growth of studies dedicated to the energy-extractive nexus (see also Bainton et al. 2021). Rogers (2015) establishes a correlation between these waves of interest and higher peaks in oil prices, and concomitant perceptions of its scarcity, which seems to guide his choice to examine the anthropological literature on oil using two analytical lenses: temporality and materiality.

To start with the second, materiality, Rogers shows how anthropologists have analysed how oil, as a material substance, and its extraction enters "social, cultural, political, and economic relationships" (ibid.: 370). He argues that the many materialities of oil currently studied (spatial, corporate, infrastructural, chemical, and microbial) have helped to advance and diversify the ways in which scholars understand representations of oil and, following the work of Appel et al. (2015b), critically scrutinize the ways in which oil serves as a metonym for some of the organizing principles of the world.

Rogers' other axis of analysis, temporality, departs from Ferry and Limbert's work (2008), which stresses that resources have distinct temporal qualities with distinct social effects. In a similar vein, Rogers approaches oil's temporality as "the ways in which the oil complex shapes senses of cyclical boom and bust, of acceleration and deceleration, and of past, present, and future" (Rogers 2015: 365). Subsequently, he discusses the densely layered and iterative dynamics of booms and busts and their associated "spectacles" and "dramas", the temporal dynamics in processes of estimating oil reserves, and those connected with financial markets and the corporate sphere. This highlights that extraction's many temporalities are complex entanglements, including aspects such as the strategies of states and corporations (among other actors), global economic conjunctures and the expectations of labourers and communities. Some of these dynamics recur so regularly that it is tempting to consider them predictable; however, anthropological studies on oil show that it has social, environmental, political, and economic effects that cannot be known once and for all. Each context produces its own specific ways of relating to this resource, meaning that "oil continues to become known in new ways, opening new possibilities as much as closing others" (ibid.: 375). It follows that anthropology, with context analysis as one of its main strengths, is well positioned to deal with this openness.

An exploding field

As indicated above, in the past two decades the anthropological scholarship on resource extraction has kept pace with the significant growth of extractive practices themselves. In the following paragraphs we restrict ourselves to pointing out a number of editorial publications that have focused on specific themes, regions, or minerals in order to outline the heterogeneity of interests of anthropologists of resource extraction and to explain our volume's rationale in relation to this heterogeneity.

Most editorial collections have concentrated on specific topics. In addition to those already mentioned in this introduction, of interest are issues related to, for example, materiality, thereby approaching the question of what resources are and how they "become" (Richardson and Weszkalnys 2014); violence or resistance, foregrounding the manifold violent processes that underpin and emanate from extraction (Shapiro and McNeish 2021; Zhouri 2017); the role of activism in responding to the challenges of extraction (Jalbert et al. 2017a); the ethics of energy dilemmas (High and Smith 2019) and the impacts and contestations around energy more broadly (Szolucha 2018); temporality, asking how conceptions and politics of time inform the processes, experiences, and anticipations around extraction (D'Angelo and Pijpers 2018; Ferry and Limbert 2008); the ways in which the corporate form is shaped by and shapes everyday life (Welker et al. 2011); the particularities of highly contested extractive technologies such as fracking (Willow and Wylie 2014); the involvement of, or particular effects, on women (Grieco and Jenkins 2021; Lahiri-Dutt and Macintyre 2006); or the role of the state, including the dialectical relationship between its presence and absence (Bainton and Skrzypek 2021).

Whereas many volumes situate their thematic approach in a cross-regional perspective, ideally enabling some degree of comparison between these different contexts, a good number of edited works have also taken in specific regional contexts – usually defined at a continental or supranational level, although at times also at national level. Such work includes analyses of mining and social transformation in Tanzania (Bryceson et al. 2013) or the idea of frontiers in sub-Saharan Africa (Werthmann and Grätz 2012); relations between mining and indigenous people in Australia and Papua New Guinea (Altman and Martin 2009; Rumsey and Weiner 2004); relations between conflict, development, and extraction in South America (Bebbington 2012); the influence of contested extractive endeavours on indigenous life-making projects in Latin America (Rivera Andía and Vindal Ødegaard 2019); the strategies used by people in the Amazon to navigate conservation and extractivism (High and Oakley 2020); research on and practices of stakeholder dialogue around oil and gas extraction in the Russian North, Siberia, and the Russian Far East (Wilson and Stammler 2006); and the making of "home" in the context of settler colonialism and environmental change in subarctic Canada (Westman et al. 2019).

In addition to topically and regionally focused works, further editorial collections have concentrated on specific resources, including coal (Lahiri-Dutt 2016), diamonds (Vlassenroot and van Bockstael 2008), gold (Panella 2010; Verbrugge and Geenen 2020); or oil and gas (Appel et al. 2015a; Logan and McNeish 2012). This material focus enables scholars to centralize how the processes, landscapes, relations, effects, and impacts (among other things) associated with extraction need to be articulated in the contexts and (material) affordances of specific resources, which direct how extraction takes shape and how it affects people, livelihoods, and environments. A very straightforward yet illustrative example of the importance of such material specificities is the issue of mercury and cyanide use and its environmental and health impacts. Whereas these are crucial concerns in the context of large and small-scale *gold* mining, they are not relevant for large and small-scale *diamond* mining, for example, simply because these substances do not play a role in diamond mining.

To conclude, building upon the excellent body of scholarship that has been produced in recent decades (editorial works, monographs, articles, and reviews), our work combines the approaches taken in review articles and editorial collections. Indeed, while annual reviews have typically been cross-topical and cross-regional in distinguishing key trends or themes in anthropological studies of resource extraction, and edited publications and special journal issues have typically focused on (a combination of) specific regions, topics, and/or materials, this volume approaches this literature by focusing on several key themes that cut across regions and materials.

5 About this volume

Aside from its aim to consolidate and further develop the anthropology of resource extraction, this volume, whose chapters are simply organized in alphabetical order, intends to function as a sort of companion, examining and indicating past and future directions taken in each of the focus themes. In this final section we provide a very brief reflection on some of our choices and explain how we see the volume's position in the larger academic field.

First, the 12 chapters in this volume are a selection of what we and the authors consider key themes in the anthropology of resource extraction, either due to their long history at the centre of studies of extraction or because they are forcefully emerging. Of course, we are far from believing that they represent *all* or the *only* key themes that one could possibly distinguish. For example, issues related to temporality, mobility, infrastructure, indigenous communities, gender, conflict and war, or the state could have been at the core of separate chapters. However, we assure the reader that, first, there are very recent literature reviews on (a good number of) these themes, including those mentioned earlier in this introduction, and second, that these aspects or themes also emerge in several of the following chapters. In developing the organizational framework of this volume, and in the process of narrowing down different thematic options, we have therefore opted for themes that would be able to cover as much ground as possible (in other words, themes that comprise but also go beyond those mentioned above), while simultaneously remaining concise and focused.

Second, although we have tried to be as comprehensive as possible, numerous interesting works have been left out, if only because our (that is, all the authors') personal interests, expertise, concerns, and language skills have informed directions taken, while the space available to each chapter has been another limiting factor for all of us. Yet this volume offers insights into a broad palate of debates and an extensive part of the anthropological literature on resource extraction.

Third, this collection sketches the contours of the anthropology of resource extraction, but does not see or set its boundaries rigidly. Like any other field, whether disciplinary or thematic, the anthropology of resource extraction did not and will not develop in isolation from insights generated elsewhere. The ways in which we engage in debates about materiality, verticality, technology, landscapes, resource curses, waste and toxicity, kinship, reciprocity, neoliberalism, and capitalism are always related to how these debates are approached in other (thematic) fields in and beyond anthropology. One must, therefore, not be surprised to read discussions involving the work of geographers, political scientists, or historians. Simultaneously, we also encourage scholars with other disciplinary backgrounds to read this volume, not least because with its in-depth attention to actual places and actual people, and its outlook on the world from below and from within, anthropology offers insights that, we believe, are distinctive to this discipline. Furthermore, the insights generated by anthropologists of resource extraction of course enter into domains that stretch beyond resource extraction, and are of relevance to anyone interested in materiality, political economy and political ecology, sustainability and development, the history of global connections and migration, gender and labour, or imaginaries and cosmology, to name but a few possibilities. In other words, this volume engages with and analyses a wide range of debates that are fundamental to dialogues among anthropologists of resource extraction and the broader anthropological and non-anthropological community.

Finally, taking both the above and the following into account, this volume, which is the first edited book dealing with the anthropology of resource extraction as a recognizable field of inquiry, could (and should) be read as an in-depth starting point, whether for those that want to acquaint themselves with the field, or those already involved. As such, we envision that this book will stimulate further reading, the further development of new ideas and the revisiting of old ones and further questioning, ultimately advancing our understanding of this phenomenon in which we are all, albeit in highly different and unequal ways, entangled: resource extraction.

Notes

1 Natural Resources Satellite Account data provided by Statistics Canada, accessible via: https://www.nrcan.gc.ca/science-data/science-research/earth-sciences/earth-sciences-re sources/earth-sciences-federal-programs/minerals-sector-employment/16739#s2

2 Minerals Council of Australia (2019). The Next Frontier, accessible via: https://treasury. gov.au/sites/default/files/2019-03/360985-Minerals-Council-of-Australia.pdf.

3 Chambers of Mines of South Africa (2016). MINE SA 2016 Facts and Figures Pocketbook, accessible via: https://www.mineralscouncil.org.za/industry-news/publications/fa cts-and-figures.

4 For references to text books that discuss anthropology as a discipline see, among many others, Barnard and Spencer (2009); Brown et al. (2020); Eriksen (2001 [1995]); Lavenda and Schultz (2013); Rapport and Overing (2002).

5 Cambridge Dictionary, available at www.dictionary.cambridge.org/dictionary/english/ extraction. Accessed 7 April 2020.

6 http://www.websters1913.com/words/Resource

7 *The Century Dictionary*, New York, NY: The Century Co., 1911.

8 Here we are talking about resources, or natural resources, in a generic sense. We are not talking about the classification of "resource" which refers to the "known" amount of a particular mineral deposit and which is used in relation to the category of "reserve", or that part of the resource which is economically feasible to mine (given a plurality of technical, legal, political, economic, and also social factors).

9 Although we include in this section some references, we want to emphasize that our reflections apply to review articles, implying that below these reflections are a large number of research articles. We therefore encourage the reader to consult these review articles and the references used.

References

Acosta A. (2013) Extractivism and neoextractivism: Two sides of the same curse. In Lang M. and Mokrani D. (eds) *Beyond development: alternative visions from Latin America*. Amsterdam; Quito: Transnational Institute; Rosa Luxemburg Foundation, 61–86.

Acosta A. (2017) Post-extractivism: From discourse to practice – reflections for action. *International Development Policy / Revue internationale de politique de développement* 9 (1): 77–101.

Altman J. and Martin D. (2009) *Power, culture, economy: Indigenous Australians and mining*, Canberra: ANU Press.

Appel H., Mason A. and Watts M. (2015a) *Subterranean estates: Life worlds of oil and gas*, Ithaca: Cornell University Press.

Appel H.C., Mason A. and Watts M. (2015b) Introduction: Oil talk. In: Appel H.C., Mason A. and Watts M. (eds) *Subterranean estates: Life worlds of oil and gas*. Ithaca: Cornell University Press, 1–26.

Arboleda M. (2020) *Planetary mine: Territories of extraction under late capitalism*, London: Verso Books.

Arsel M., Hogenboom B. and Pellegrini L. (2016) The extractive imperative in Latin America. *The Extractive Industries and Society* 3 (4): 880–887.

Bainton N. (2020) Mining and Indigenous Peoples. *Oxford Research Encyclopedia of Anthropology*, Oxford: Oxford University Press.

Bainton N., Kemp D., Lèbre E., et al. (2021) The energy-extractives nexus and the just transition. *Sustainable Development* 29 (4): 624–634.

Bainton N. and Skrzypek E. (2021) *An absent presence: Encountering the state through natural resource extraction in Papua New Guinea and Australia*. Canberra: ANU Press.

Ballard C. and Banks G. (2003) Resource wars: The anthropology of mining. *Annual Review of Anthropology* 32 (1): 287–313.

Barnard A. and Spencer J. (2009) *The Routledge encyclopedia of social and cultural anthropology*. London & New York: Routledge.

Bebbington A. (2012) *Social conflict, economic development and extractive industry: Evidence from South America*, London: Routledge.

Behrends A., Reyna S. and Schlee G. (2011) *Crude domination: An anthropology of oil*, New York & Oxford: Berghahn Books.

Boivin N. (2004) From veneration to exploitation: Human engagement with the mineral world. In: Boivin N. and Owoc M.A. (eds) *Soils, stones and symbols: Cultural perceptions of the mineral world*. London: UCL Press, 1–30.

Boivin N. and Owoc M.A. (2004) *Soils, stones and symbols: Cultural perceptions of the mineral world*. London: UCL Press.

Bonneuil C. and Fressoz J-B. (2016) *The shock of the Anthropocene: The earth, history and us*, London & New York: Verso Books.

Boyer D. (2014) Energopower: An Introduction. *Anthropological Quarterly* 87: 309–333.

Brand U., Dietz C. and Lang M. (2016) Neo-Extractivism in Latin America – one side of a new phase of global capitalist dynamics. *Ciencia Política* 11 (21): 125–159.

Bridge G. (2004) Contested terrains: Mining and the environment. *Annual Review of Environment and Resources* 29 (1): 205–259.

Bridge G. (2009) Material worlds: Natural resources, resource geography and the material economy. *Geography Compass* 3 (3): 1217–1244.

Brown N., McIlwraith T. and de González L.T. (2020) *Perspectives: An open introduction to cultural anthropology*. American Anthropological Association, http://perspectives.americana nthro.org

Bryceson D.F., Fisher E., Jønsson J.B., et al. (2013) *Mining and social transformation in Africa: Mineralizing and democratizing trends in artisanal production*, London & New York: Routledge.

Bunker S.G. (1984) Modes of extraction, unequal exchange, and the progressive under-development of an extreme periphery: The Brazilian Amazon, 1600–1980. *American Journal of Sociology* 89 (5): 1017–1064.

Burton M., Biesheuvel T. and Longley A. (2021) *When does a commodities boom turn into a supercycle?* Bloomberg, 26 February. https://www.bloomberg.com/news/articles/2021-02-26/when-does-a-commodities-boom-turn-into-a-supercycle-quicktake. Accessed 7 April 2021.

Buss D. and Rutherford B. (2020) Gendering women's livelihoods in artisanal and small-scale mining: An introduction. *Canadian Journal of African Studies / Revue Canadienne des Études Africaines* 54 (1): 1–16.

Ciccantel P. and Smith D.A. (2009). Rethinking global commodity chains. Integrating extraction, transport, and manufacturing. *International Journal of Comparative Sociology* 50 (3–4): 361–384.

Cremers L., Kolen J. and de Theije M.E.M. (2013) *Small-scale gold mining in the Amazon. The cases of Bolivia, Brazil, Colombia, Peru and Suriname*. Cuadernos del CEDLA 26, CEDLA.

Cuvelier J., Vlassenroot K. and Olin N. (2014) Resources, conflict and governance: A critical review. *The Extractive Industries and Society* 1 (2): 340–350.

D'Angelo L. and Pijpers R.J. (2018) Mining temporalities: An overview. *The Extractive Industries and Society* 5 (2): 215–222.

Dolan C. and Rajak D. (2016) *The anthropology of corporate social responsibility*, New York: Berghahn Books.

Downing T.E. (2002) *Avoiding new poverty: Mining-induced displacement and resettlement*, London: International Institute for Environment and Development.

Elhacham E., Ben-Uri L., Grozovski J., et al. (2020) Global human-made mass exceeds all living biomass. *Nature* 588 (7838): 442–444.

Epstein A.L. (1958) *Politics in an urban African community*, Manchester: Manchester University Press.

Ericsson M. and Löf O. (2019) Mining's contribution to national economies between 1996 and 2016. *Mineral Economics* 32 (2): 223–250.

Eriksen T.H. (2001 [1995]) *Small places, large issues*, London: Pluto Press.

Eriksen T.H. (2016) *Overheating: Coming to terms with accelerated change*. London: Pluto Press.

Falk-Moore S. (1994) *Anthropology and Africa: Changing perspectives on a changing scene*, Charlottesville & London: The University Press of Virginia.

Ferry E.E. and Limbert M.E. (2008) *Timely assets*, Santa Fe: School for Advanced Research Press.

Ferry E., Vallard A. and Walsh A. (2019) *The anthropology of precious minerals*, Toronto: University of Toronto Press.

Gilberthorpe E. and Rajak D. (2016) The anthropology of extraction: Critical respersctives on the resource curse. *The Journal of Development Studies* 53 (2): 186–204.

Gluckman M. (1961) Anthropological problems arising from the African industrial revolution. In: Southall A. (ed.) *Social change in modern Africa*. London: Oxford University Press.

Godoy R. (1985) Mining: Anthropological perspectives. *Annual Review of Anthropology* 14 (1): 199–217.

Golub A. (2019) Mining. *Cambridge encyclopedia of anthropology*. Cambridge: Cambridge University Press.

Gómez-Barris M. (2017) *The extractive zone: Social ecologies and decolonial perspectives*, Durham: Duke University Press.

Grieco K. and Jenkins K. (2021) Introduction: Articulating gender and resource extraction in Latin America. *Bulletin of Latin American Research* 40: 169–171.

Gudynas E. (2010) The new extractivism of the 21st century: Ten urgent theses about extractivism in relation to current South American progressivism. *Americas Program Report* 21: 1–14.

Gudynas E. (2014) Buen Vivir: sobre secuestros, domesticaciones, rescates y alternativas. In: Oviedo Freire A. (ed.) *Bifurcación del Buen Vivir y el sumak kawsay*. Quito: Ediciones SUMAK, 23–45.

High C. and Oakley R.E. (2020) Conserving and extracting nature: Environmental politics and livelihoods in the new "middle grounds" of Amazonia. *The Journal of Latin American and Caribbean Anthropology* 25: 236–247.

High M.M. and Smith J.M. (2019) Introduction: The ethical constitution of energy dilemmas. *Journal of the Royal Anthropological Institute* 25 (S1): 9–28.

Hilson G. (2002) Small-scale mining in Africa: Tackling pressing environmental problems with improved strategy. *The Journal of Environment & Development* 11: 149–174.

Hilson G. and McQuilken J. (2014) Four decades of support for artisanal and small-scale mining in sub-Saharan Africa: A critical review. *The Extractive Industries and Society* 1 (1): 104–118.

Huber M.T. (2017) Petrocapitalism. *International Encyclopedia of Geography*, Wiley online, 1–6.

ICMM (2020) *Role of mining in national economies: Mining Contribution Index (MCI)*. London: The International Council on Mining and Metals.

IEA (2021) *The role of critical minerals in clean energy transitions*. Paris: International Energy Agency.

IIED (2017) *Global trends in Artisanal and Small-Scale Mining (ASM): A review of key numbers and issues*. Winnipeg: International Institute for Environment and Development.

Jacka J.K. (2018) The anthropology of mining: The social and environmental impacts of resource extraction in the mineral age. *Annual Review of Anthropology* 47 (1): 61–77.

Jalbert K., Willow A., Casagrande D., et al. (2017a) *ExtrACTION: Impacts, engagements, and alternative futures*, New York: Routledge.

Jalbert K., Willow A., Casagrande D., et al. (2017b) Introduction: Confronting extraction, taking action. In: Jalbert K., Willow A., Casagrande D., et al. (eds) *ExtrACTION: Impacts, engagements, and alternative futures*. New York: Routledge, 1–13.

Kirsch S. (2010) Sustainable mining. *Dialectical Anthropology* 34 (1): 87–93.

Knapp A.B., Pigott V.C. and Herbert E.W. (2002) *Social approaches to an industrial past: The archaeology and anthropology of mining*. London & New York: Routledge.

Labban M. (2014) Deterritorializing extraction: Bioaccumulation and the planetary mine. *Annals of the Association of American Geographers* 104 (3): 560–576.

Lahiri-Dutt K. (2016) *The coal nation: Histories, ecologies and politics of coal in India*, London: Routledge.

Lahiri-Dutt K. (2018) *Between the plough and the pick: Informal, artisanal and small-scale mining in the contemporary world*, Canberra: ANU Press.

Lahiri-Dutt K. and Macintyre M. (2006) *Women miners in developing countries: Pit women and others*, Ashgate: Aldershot.

Lavenda R.H. and Schultz E.A. (2013) *Anthropology: What does it mean to be human?*, New York: Oxford University Press.

Logan O. and McNeish J-A. (2012) *Flammable societies: Studies on the socio-economics of oil and gas*, London: Pluto Press.

Marcus G.E. (1995) Ethnography in/of the world system: The emergence of multi-sited ethnography. *Annual Review of Anthropology* 24 (1): 95–117.

Mathews G., Ribeiro G.L. and Vega C.A. (2012) *Globalization from below: The world's other economy*, London & New York: Routledge.

Maus V., Giljum S., Gutschlhofer J., et al. (2020) A global-scale data set of mining areas. *Scientific Data* 7 (1): 289.

Mezzadra S. and Neilson B. (2017) On the multiple frontiers of extraction: Excavating contemporary capitalism. *Cultural studies* 31 (2–3): 185–204.

Mitchell J.C. (1956) The Kalela dance: Aspects of social relationship among urban Africans in Northern Rhodesia. *Rhodes-Livingstone Paper* 27: 233.

Mitchell T. (2011) *Carbon democracy: Political power in the age of oil*, London & New York: Verso Books.

Mollona M. (2019) *Brazilian steel town: Machines, land, money and commoning in the making of the working class*, New York & Oxford: Berghahn Books.

Nash J.C. ([1979] 1993) *We eat the mines and the mines eat us: Dependency and exploitation in Bolivian tin mines*, New York: Columbia University Press.

Oberle B., Bringezu S., Hatfield-Dodds S., et al. (2019) *Global resources outlook 2019: Natural resources for the future we want*. A Report of the International Resource Panel (IRP). International Resource Panel. United Nations Environment Programme.

OECD (2019) *Global material resources outlook to 2060*. https://www.oecd.org/env/global-material-resources-outlook-to-2060-9789264307452-en.htm.

Panella C. (2010) *Worlds of debts: Interdisciplinary perspectives on gold mining in West Africa*, Amsterdam: Rozenberg.

Parks J. (2021) The poetics of extractivism and the politics of visibility. *Textual Practice* 35 (3): 353–362.

Perlez J. and Johnson K. (2010) Behind gold's glitter: Torn lands and pointed questions. *The New York Times*. Accessed 8 February 2020.

Pijpers R.J. and Eriksen T.H. (2018) *Mining encounters: Extractive industries in an overheated world*, London: Pluto Press.

Rapport N. and Overing J. (2002) *Social and cultural anthropology: The key concepts*, London & New York: Routledge.

Richardson T. and Weszkalnys G. (2014) Introduction: Resource materialities. *Anthropological Quarterly* 87 (1): 5–30.

Rivera Andía J.J. and Vindal Ødegaard C. (2019) *Indigenous life projects and extractivism: Ethnographies from South America*, Cham: Springer International Publishing.

Rogers D. (2015) Oil and anthropology. *Annual Review of Anthropology* 44 (1): 365–380.

Rumsey A. and Weiner J.F. (2004) *Mining and indigenous lifeworlds in Australia and Papua New Guinea*, Wantage: Sean Kingston Publishing.

Shapiro J. and McNeish J-A. (2021) *Our extractive age: Expressions of violence and resistance*, London: Routledge.

Smith J. and High M.M. (2017) Exploring the anthropology of energy: Ethnography, energy and ethics. *Energy Research & Social Science* 30: 1–6.

Smith J.M. and Kirsch S. (2018) Mining. In: Callan H. (ed.) *The international encyclopedia of anthropology*. Wiley online, 1–4.

Sonter L.J., Dade M.C., Watson J.E., et al. (2020) Renewable energy production will exacerbate mining threats to biodiversity. *Nature Communications* 11 (4174): 1–6.

Steffen W., Crutzen P. and McNeill J. (2007) The Anthropocene: Are humans now over-whelming the great forces of nature? *AMBIO: A Journal of the Human Environment* 36: 614–621.

Svampa M. (2019) *Neo-extractivism in Latin America: Socio-environmental conflicts, the territorial turn, and new political narratives*, Cambridge: Cambridge University Press.

Szeman I. and Wenzel J. (2021) What do we talk about when we talk about extractivism? *Textual Practice* 35: 505–523.

Szolucha A. (2018) *Energy, resource extraction and society: Impacts and contested futures*, London: Routledge.

Taussig M. (1980) *The devil and commodity fetishism in South America*, Chapel Hill: University North Carolina Press.

Temper L., Del Bene D. and Martinez-Alier J. (2015) Mapping the frontiers and front lines of global environmental justice: the EJAtlas. *Journal of Political Ecology* 22 (1): 255–278.

Verbrugge B. and Geenen S. (2020) *Global gold production touching ground: Expansion, informalization, and technological innovation*, Cham: Palgrave Macmillan.

Vlassenroot K. and van Bockstael S. (2008) *Artisanal diamond mining: Perspectives and challenges*, Ghent: Academia Press.

Welker M., Partridge D.J. and Hardin R. (2011) Corporate lives: New perspectives on the social life of the corporate form: An introduction to supplement 3. *Current Anthropology* 52 (S3): 3–16.

Werthmann K. and Grätz T. (2012) *Mining frontiers. Anthropological and historical perspectives*, Cologne: Rüdiger Köppe Verlag.

Westman C.N., Joly T.L. and Gross L. (2019) *Extracting home in the oil sands: Settler colonialism and environmental change in subarctic Canada*, London: Routledge.

Willow A.J. and Wylie S. (2014) Politics, ecology, and the new anthropology of energy: Exploring the emerging frontiers of hydraulic fracking. *Journal of Political Ecology* 2 (1): 222–236.

Wilson E. and Stammler F. (2006). The oil and gas industry, local communities and the state. *Sibirica: Interdisciplinary Journal of Siberian Studies* 5 (2).

Wilson G. (1941) *An essay on the economics of detribalization in Northern Rhodesia*. Rhodes-Livingstone Papers. Livingstone, Northern Rhodesia: Rhodes-Livingstone Institute.

Zhouri A. (2017) Introduction: Anthropology and knowledge production in a "minefield". *Vibrant: Virtual Brazilian Anthropology* 14 (2).

2

CORPORATIONS

Elana Shever

1 Introduction

There is no doubt that extractive corporations have powerful effects on the contemporary world. Anthropological scholarship shows that this power does not
simply emanate from these companies' financial might, but from the complex webs
of people, materials, capital, technology, and knowledge that they weave together.
For almost a century, anthropologists have documented the consequences of corporate extraction for people around the world and the environments in which they
live. This chapter begins by examining the early scholarship, which focuses on
labour migration to and from mining camps, and the later shift towards ethnographies of mining enclaves. Here, I highlight the importance of corporate
paternalism in shaping the lives of workers and their families. The second section of
the chapter turns to ethnographic studies of extractive corporations' role in globalization, particularly in instituting its neoliberal forms. The third section scrutinizes
"corporate social responsibility" as a key feature of neoliberal capitalism in the
extractive industries. In the penultimate section, I explore recent analyses of the
legal definition of the corporate form that have reconceptualized extractive corporations as potent *effects* of capitalist processes, rather than as invincible *agents* of
capitalism. Throughout these shifts in foci and theoretical lens, anthropological
scholarship has continued to reveal the many ways in which extractive corporations
both shape their social and material environments and are shaped by them.

2 Labour, enclaves, and paternalism

From hunting for gold in the Americas to prospecting for copper in Oceania, the
search for subsurface resources has led corporate agents to venture into the same
places where anthropologists have travelled since the beginning of the twentieth

DOI: 10.4324/9781003018018-2

century. In these ostensibly isolated locations, company representatives come into contact with the indigenous people among whom anthropologists traditionally studied. Yet the majority of early anthropologists excluded extractive operations from their work in an attempt to capture ways of life before the arrival of outsiders. Anthropologists associated with the Manchester School in the 1930s and 1940s broke away from the standard of portraying "untouched natives" to depict how colonial mining operations produced social change across Central and Southern Africa. Richards's ([1939] 1995) study of agriculture and nutrition among the Bemba people of Zambia (then Rhodesia) is exemplary. Although Richards did not address mining explicitly, she nonetheless documented how the migration of Bemba men to mining camps, where they worked for up to three years at a time, disrupted everything from food production, distribution, and consumption, to kinship, gender, and political organization in Bemba labour-sending communities. Foreign mining firms used either their own strong-armed recruitment tactics or colonial authorities to supply forced labourers for the gruelling labour of resource extraction, while relying on labour-sending communities for the maintenance and social reproduction of their workforce (Godoy 1985: 206). Despite this, "the smokestacks and smelters of the mines … symbolized everything that was understood to be industrial and modern" (Ferguson 1999: 35) to many in Africa, including the anthropologists of the Manchester School.

While most of the first anthropologists to examine social change in Africa extolled mining companies as agents of modernization, subsequent ones demonstrated the negative consequences of corporate extraction. Studies from Latin America and elsewhere joined those from Africa in detailing how mining pushed people deeper into the highly unequal and unstable relationships of wage labour, commodity consumption, and capital (Godoy 1985: 206–207). Miner communities and companies remained on the sidelines of this scholarship; thus Godoy's 1985 review of the Anthropology of Mining could accurately state that the discipline had only "recently discovered" miners (199). Nash's *We Eat the Mines and the Mines Eat Us* ([1979] 1993), an ethnography of two Bolivian tin mines, was a harbinger of the scholarship to come. Yet it would take another decade for Anthropology to "discover" extractive corporations as an equally fitting topic for ethnographic research. In early ethnographies of mining communities, "the figure of 'the mining company' lurks monolithically and often menacingly in the background of many anthropological accounts of communities affected by mining operations" (Ballard and Banks 2003: 290). Nash's ethnography, for example, moves between a sympathetic portrait of the struggles of miners and their families and a stinging critique of capitalism, while skipping over the corporation that mediated between worker families and the global tin industry. When mining firms are mentioned, they are often portrayed as personifications of their distant, heartless owners. "Corporations, from this perspective, are seen as tools of a capitalist class and accordingly stand in opposition to the working class" (Urban and Koh 2012: 142). Twentieth-century ethnographies of mining communities thus include critiques of extractive corporations, but not ethnographic research about the corporations themselves.

The "discovery" of miners as an important subject for Anthropology never-theless led to a rich body of scholarship on mining communities. Studies from the Americas (e.g. Tinker Salas 2009), Africa (e.g. Ferguson 1999), and the Middle East (e.g. Vitalis 2007) have amplified Nash's analysis of how extractive corporations reshape the social, political, economic, and physical landscapes of the places in which they operate. Anthropologists have tended to regard these sites as a parti-cular form of enclave, in which a foreign corporation insulates its operations from its national context, essentially creating a state within a state (e.g. Coronil 1997: 108–109). As extraction expanded around the world in the twentieth century, companies usually recruited people from nearby to carry out the dangerous labour of extraction in exchange for cash that these workers would not otherwise earn, while bringing in outsiders for technical and managerial positions. The firms introduced infrastructure such as piped water, technologies such as electricity, and services, from hospitals to theatres, which were previously unimaginable in the region (e.g. Finn 1998). To this day, corporations restrict access to these resources to the privileged few within their enclaves, and deploy them as measures of rank.

Enclaves are designed to isolate extraction – its corporate sponsors in particular – from political instability, social unrest, and economic demands from the surround-ing society. Many are fortified with walls (e.g. Ferguson 2005), policed by private security forces (e.g. Zalik 2004), or protected by paramilitary apparatuses (e.g. Chowdhury 2016). Others are secured through "soft" technologies, such as com-munity development programmes (e.g. Welker 2014). Yet they are never isolated from the outside world. As Ferguson put it, "enclaves of mineral-extractive investments … are normally tightly integrated with the head offices of multi-national corporations and metropolitan centres, but sharply walled off from their own national societies" (2005: 379). Despite their walls, extractive enclaves are, in fact, entangled with national societies as well as with foreign metropoles (Appel 2019).

As extractive corporations have attempted to separate their operations from their national contexts, they have also segregated the people within them. Company executives have established not only distinct levels of compensation among their employees, but also differentiated housing and myriad other benefits. The inequality within company enclaves is ostensibly based on skill levels and job categories, but has historically mirrored the racial, ethnic, national, and gender discrimination of the corporation's country of origin (Appel 2019; Porteous 1974; Vitalis 2007). The US oil firm ARAMCO exemplified "the long, unbroken legacy of hierarchy across the world's mineral frontiers" when it re-created US-style (Jim Crow) racial segregation within its enclaves in Saudi Arabia during the 1930s (Vitalis 2007: 18). Segregation has been so stark that the mere sight of a few Fili-pinos in the gym reserved for upper management "creeped out" a white North American woman in an oil company compound in Equatorial Guinea (Appel 2019: 91–92). These social divisions have impeded labour solidarity and thus unionization. Yet extractive enclaves have long and bitter histories of strikes and, concomitantly, of their violent suppression by corporate and state forces (e.g.

Dinius and Vergara 2011). At least in some cases (e.g. Yessenova 2012), the enclaving of extraction has fostered the very threats it was designed to prevent.

Despite the dearth of ethnographic attention paid to extractive corporations themselves until recently, anthropologists have given significant attention to corporate paternalism. While paternalism was widespread across many industries in the twentieth century (e.g. Dinius and Vergara 2011), it has been a particularly strong and enduring force in extraction. In numerous instances, extractive corporations replaced the forced labour and punitive discipline of colonialism with a familial mode in which the company supported workers and their kin in exchange for their hard labour. As one woman told Nash, "my husband lends his lungs to the company and they lend us this house" ([1979] 1993: 94). Paternalism frames the relationship between employer and employee in terms of kinship obligation and reciprocal care. Yet it encompasses far more than discourse. It also entails corporate management of the minutiae of people's lives. Paternalistic companies have controlled workers' movements not only in and out of mining enclaves, but even within their homes. They have prevented workers from quitting by delaying their pay and encouraging them to accumulate debt to the company store. Extractive corporations have more subtly regulated the lives of workers and their families through biopolitical means, such as the provision of healthcare, education, and recreation. In the early 2000s, for instance, the multinational mining giant Anglo American started providing HIV treatment to its workers in South Africa, in the context of soaring HIV rates and a state healthcare system unable to deal with the pandemic (Rajak 2016). HIV+ miners thus came to depend on the companies not only for their livelihoods but also for bare life. "As welfare is conflated with the maintenance of human capital," Rajak explains, "care becomes a conduit for corporate control, giving new force to old paternalistic and repressive regimes of labour management" (2016: 37). Although corporate paternalism invokes reciprocity, it establishes rigid hierarchies and regulates relationships, from conjugal to national ones.

How much of life can a corporation determine? In the "model" company town of Kleinzee, South Africa, the diamond industry giant De Beers controlled nearly every aspect of life – often through brokers who tricked and manipulated miners – for the better part of the twentieth century. "Everything associated with the settlement – the houses, shops, schools, hospitals, and recreation facilities – was subordinated to the business enterprise and the will of its owners" (Carstens 2001: 3). Carstens asserts that De Beers was able to maintain the allegiance of its employees, despite its abuse of them, through an efficacious combination of coercion and persuasion, which he calls "obligated loyalty" (Carstens 2001: 186–187). De Beers was so successful at controlling affairs at the regional, national, and international levels that Carstens's portrait of it as omnipotent may be justified. Yet other ethnographies describe how extractive corporations' efforts to manipulate their workforces have backfired. Company stores provide a lucid example. In many cases, extractive corporations have operated the only store in town, charging high prices and offering loans that lead worker families to become endlessly indebted to them.

Nevertheless, these stores have also served as meeting places where women discuss their problems and organize strikes against the firm (e.g. Finn 1998: 98–99; Nash [1979] 1993: 91–97). Paternalistic attempts to organize workers' kinship have had similarly unpredictable results. In Zambian copper mines, for instance, a Christian model of the nuclear family did not take firm hold, despite mining companies' promotion of it through educational campaigns, compensation schemes, and other supports (Ferguson 1999; cf. Rubbers 2015). Furthermore, extractive corporations have been unable to jettison paternalistic obligations to worker families once they no longer serve their goals.

Argentina's state oil company, Yacimientos Petrolíferos Fiscales (YPF) illustrates the continued ambivalence of paternalism as a tool of corporate discipline. YPF used paternalistic practices to build a dedicated and docile workforce in the oil fields of Patagonia in the early twentieth century (Shever 2012). Its workers came to identify so strongly with the company that they called themselves "YPFianos". When YPF was privatized at the end of the century, the vast majority of YPFianos lost their jobs but sustained the expectation that the company would provide them with lifelong support. With the backing of the state, many YPFianos became worker–owners in microenterprises (*emprendimientos*) that demanded contracts with the privatized YPF. These contracting arrangements drew the company back into the very paternalistic relationships that the privatization was supposed to eliminate. Corporate paternalism, therefore, is not merely an antecedent to the neoliberal corporate social responsibility discussed below, but has coexisted ambivalently with it (Shever 2013; see also Billo 2015).

With the flourishing of the Anthropology of Extraction in the late twentieth century, extractive corporations began to come out of the ethnographic shadows. Carstens's long-term scholarship in South Africa exemplifies how Anthropology shifted from studying labour-sending communities (1966) to company towns (2001). While ethnographies of mining- and oil-worker communities reveal a great deal about how corporations shape the social dynamics of extractive enclaves, they do not account fully for the transnational character of extraction. Finn's comparative investigation of the Anaconda Company's mines in the United States and Chile leads her to assert that "it is impossible to grasp the cultural, historical, political, and social implications of copper production in one locale without viewing it in relation to the other," as well as "the cultural politics of copper writ large" (1998: 12). More than a decade after Nash (1979) called for a multiscalar Anthropology of Transnational Corporations, Finn employed "multisited ethnography" (Marcus 1995) to examine a mining multinational. Rajak's (2011) study of the mining giant Anglo American expanded Finn's precedent by conducting fieldwork in London, where Anglo American is headquartered; in Johannesburg, the symbolic home of the company; and in Rustenburg, where it extracts platinum. By the turn of the twenty-first century, anthropologists were exploring various transnational mining companies in global perspective. I now turn to those who have examined extractive corporations as a crucial component of global capitalism and, specifically, have shown how these enterprises have been instrumental in the emergence of its neoliberal form.

3 Globalization and the rise of neoliberal capitalism

By the turn of the twenty-first century, the anthropological study of the social dynamics of mining communities had given way to the examination of extraction as part of global capitalism. Anthropologists' attention to the details of everyday life has revealed how capitalist processes play out differently for different people. "As institutions that pervade the social and material fabric of everyday life, corporations shape human experience not only in spectacular and disastrous ways but also in mundane, everyday, ambivalent, and positive ways" (Welker et al. 2011: 3–4). However, anthropologists have faced numerous difficulties when trying to conduct fieldwork inside corporations (see also Bainton and Skrzypek this volume). Most corporate actors have refused to "expose themselves directly to ethnographic scrutiny" (Ballard and Banks 2003: 290). Yet anthropologists have not turned away from ethnographic research; instead, they have developed innovative techniques to research extractive corporations, including participating in industry conferences and trade shows (e.g. Ferry 2013), attending corporate–community negotiations (e.g. Golub 2014) and court proceedings (e.g. Sawyer 2006), working with sub-contractors (e.g. Shever 2012) and consultants (e.g. Fortun 2001), and even taking walks with industry insiders (High 2019). Anthropology no longer leaves the investigation of corporations to other social sciences, like it once did (Gusterson 1997: 114–115; Welker et al. 2011: 5).

Recent scholarship has demonstrated the great extent to which extractive corporations shape domains far beyond their physical footprints. Anthropological analyses of speculation (e.g. Chowdhury 2016; Weszkalnys 2015), scientific research (e.g. Kirsch 2020), future forecasting (e.g. Davis-Floyd 1998; Mason 2013), advertising (e.g. Fortun 2001; Sawyer 2010), and public relations (e.g. Rajak 2014; Shever 2010) illustrate the wide range of practices in which extractive corporations are engaged. Moreover, anthropologists and allied scholars have explored how corporate-sponsored art (e.g. Barrett and Worden 2014) and public displays of oil wealth (e.g. Apter 2005; Rogers 2015) celebrate hydrocarbons and naturalize their pervasiveness. Empirical studies of the quotidian consumption of cooking fuel (e.g. Shever 2012), gasoline (e.g. Huber 2013), and diamonds (e.g. Falls 2014; Proctor 2001) illustrate the ways that extractive corporations influence the aspirations and everyday activities of people around the globe. This growing body of scholarship uncovers the multitude of subtle and not-so-subtle ways in which extractive corporations are shaping sociality and subjectivities the world over.

The expanding influence of extractive corporations needs to be understood within the context of the rise of neoliberal capitalism. Neoliberalization is predominantly defined as the execution of a series of ideologically aligned economic policies that include the transfer of state enterprises and services to private firms, a decrease in government regulation, and a shift from state to foreign investment in development. As Harvey's influential description states, "the corporatization, commodification, and privatization of hitherto public assets has been a signal feature of the neoliberal project" (2005: 160). States have ceased extracting resources themselves and, instead, have focused on facilitating private profit over providing public goods. They thus have helped corporations to expand their roles in production, distribution, and sales of hydrocarbons and minerals. Moreover, private firms have taken over the state's roles in the provision of security (e.g. Zalik 2004), healthcare, and education (e.g. Rajak 2016) in extractive zones. The rise of neoliberal capitalism has further entailed the unregulated growth of extractive corporations through mergers and acquisitions. The oil industry, for one, has coalesced into a handful of transnational "supermajors", and few projects being developed today involve only one of these.[1] A significant part of the power of extractive corporations in the contemporary world therefore derives from the enormous scale and scope of the activities that they direct.

As extractive corporations have increased the quantity of resources and wealth they control, they have concurrently decreased the size of their workforces. Their transition from labour-intensive extraction to technology- and capital-intensive extraction has meant that they no longer are filling their former role as major employers in "underdeveloped" regions. Now they hire predominantly contingent labour for the manual work that they still need, as Rubber's chapter in this volume discusses. Labour-contracting arrangements also transfer the risks of extraction from multinationals to the business entities that Sawyer (2001) has called, aptly, "corporate prostheses". The construction of the Chad–Cameroon pipeline illustrates these interconnected changes in how energy development projects are carried out within a neoliberal capitalist regime (Guyer 2011; Muñoz and Burnham 2016). This pipeline project also demonstrates how extractive corporations have shifted from directly producing resources themselves to managing complex networks of subsidiaries, contractors, and subcontractors who do the dirty work of prospecting, extracting, refining, and distributing subsurface resources, as well as assuming their risks.

Anthropologists have drawn attention to the failures of neoliberal "reforms" to live up to their promises of improving people's lives; instead, these changes have decreased wellbeing while increasing insecurity and inequality (e.g. Comaroff and Comaroff 2001). Anthropological research also reveals how neoliberal capitalism has altered far more than how natural resources are produced and distributed. It has remade the people who produce, distribute, and consume them. As Sawyer's ethnography of oil extraction in the Ecuadorian Amazon illustrates, "these new governing methods strive to shape and direct individuals to be autonomous liberal subjects who will espouse the rational economies of competition, accountability, and self-actualization" (2004: 14–15). Through their public-relations managers, community liaisons, and consultants, extractive corporations have worked to make the subjectivities of people – from corporate executives (e.g. High 2019) to subcontracted workers (e.g. Shever 2008) – fit the neoliberal ideal. Yet these efforts are not always successful. The neoliberal transformation of extraction has provoked forceful counter-movements (e.g. Li 2015; see also Li and Velásquez this volume).

Anthropologists have also returned their attention to the indigenous people examined in the classic ethnographies discussed early in this chapter, but without their predecessors' blindfolds to colonialism and capitalism. Recent ethnographies of indigenous groups living near oil fields and mineral deposits portray the diversity and complexity of their relationships with extractive corporations from a global perspective. Jacka's (2015) exploration of Porgeran communities residing downstream from gold mines in Papua New Guinea, for instance, shows both the devastation brought by mining companies and Porgerans' creativity in dealing with them. These indigenous people are certainly not the passive victims of extraction portrayed in many earlier ethnographies. Indigenous and other activists have developed global networks to hold corporations accountable for environmental destruction and ill health, to resist corporate control over land and resources, and to contest the distribution of the wealth generated by extraction (e.g. Kirsch 2014b). Although Benson and Kirsch's claim that corporations' responses to their critics are "the defining feature of contemporary capitalism" (2010: 474) may be hyperbolic, extractive corporations have not been able to ignore their critics either. The scholarship on anti-corporate activism thus provides a crucial counterview to studies depicting extractive firms' growing influence.

4 Corporate social responsibility

Extractive corporations have long attempted to manage their relationships with indigenous peoples, state agents, and other citizens through gifting (Cross 2014; Dolan and Rajak 2016). Beginning in the 1990s, these efforts crystallized in the "corporate social responsibility" (CSR) movement. Its promoters argue that while corporate charity is an expendable "add-on" divorced from core business goals, CSR is a novel "win–win" strategy that both produces benefits for society and generates financial rewards for shareholders. In the face of mounting concerns about the environmental and social harms caused by extraction, multinationals have implemented self-regulation, monitoring, and reporting (Garsten 2012). Advocates of CSR have worked to unify the world's largest corporations around voluntary principles of "corporate citizenship" and "environmental stewardship", while eschewing regulation (Ruggie 2002; Watts 2005). In the process, CSR has burgeoned into a booming transnational industry in and of itself (Dolan and Rajak 2016: 2). Critics, both inside and outside academia, have characterized CSR as a public-relations strategy to ameliorate the concerns of states, consumers, and investors without making significant changes to corporate operations (e.g. Watts 2005: 394–395). One anthropologist has likewise asserted that CSR is "always initiated, primed, and skewed in the interests of capital" (Cross 2014: 125). Yet ethnographies of particular CSR projects (including Cross 2011) tell a more nuanced story.

Corporations in numerous sectors have developed CSR programmes, yet anthropological research on CSR has focused especially on the extractive industries. By moving between corporate centres of calculation in metropoles, and extraction sites far from them, this scholarship offers critical insight into how CSR travels the globe yet is embedded in the particularities of each locale in which it is instituted (e.g. Rajak 2011). Anthropologists have demonstrated that CSR responds to neoliberal demands for corporate efficiency, quantification, and accountability, but, like other neoliberal endeavours, it has not fulfilled its promise to improve the lives of impoverished people. CSR more often leads to "exclusion, precarity, or disempowerment" for people living in extractive zones (Dolan and Rajak 2016: 4). The reformulation of corporate paternalism and philanthropy programmes into CSR has also spawned an "ethic of detachment" (Cross 2011). Extractive corporations have used CSR to win social license to operate while deflecting responsibility for the ill health and social suffering their operations cause (e.g. Coumans 2011). While some scholars assert that CSR benefits wealthy shareholders, others argue that it has not consistently advanced corporate interests (e.g. Sydow 2016; Welker 2014). Whether or not CSR has increased corporate profits, it has reinforced the neoliberal dictum that the market is the ultimate arbiter of all goods – material, social, and moral (Garsten 2012). The CSR movement has thereby advanced neoliberal capitalism in extractive zones and elsewhere.

Several CSR projects run by oil companies in South America illustrate how the practice of CSR has done more than simply maintaining corporate control in the extractive industries. Following the privatization of the oil industry in Argentina, for example, Shell faced both criticism from the president and his allies for being a foreign "market dictator" and demands for support from the impoverished people living near its refinery in Dock Sud. Shell officials responded by replacing the corporation's ad hoc donations with a structured CSR programme called "Creating Bonds". It was aimed to remake angry Dock Sud residents into the company's "community partners" and Shell into a compassionate neighbour. The employee in charge of Creating Bonds told me that, through this project, "Shell is not a brand; it's a face, a person" (Shever 2010: 35). While trying to establish a personal relationship with Dock Sud residents, she and her staff pushed them to rely less on the company to treat the illnesses caused by pollution from the refineries, and, instead, to address these problems independently. To access Creating Bonds funding, one social movement active in the area was compelled to reshape its demand that Shell clean up its operations and hire local residents, replacing it with a constrained request for money to purchase building materials and medical equipment for its members to use in providing nutrition education and basic health services to themselves, by themselves. Although this "community–corporate partnership" ultimately fell apart, it nonetheless enabled Shell to promote itself as addressing the problems in Dock Sud without acknowledging its role in creating them. Likewise, in neighbouring Chile, a CSR project involving an Atacamanian group and a mining giant drew on an indigenous conception of partnership, yet placed greater risks on the Atacamanian partners than on the corporate ones (Babidge 2013). Rajak (2008, 2011) provides further elaboration of how CSR positions corporations not only as rights-bearing citizens but also as moral persons who care about, but bear no responsibility for, other people.

5 From the corporate person to the corporate effect

The prevailing view of extractive corporations in Anthropology has never been a rosy one. Even before the rise of neoliberal capitalism, mining and oil companies were identified as the agents, if not also the causes, of immeasurable harms in the world. There is a large body of scholarship, some of it already cited in this chapter, recording how extractive corporations have increased inequality between world regions, social groups, and individuals; perpetrated violence and instigated wars; and destroyed environments and advanced global climate change (see also Bebbington and Bury 2013; Evans et al. 2002; Jacka 2018). Both human suffering and environmental destruction, Benson and Kirsch point out, are "part and parcel of [the] normal functioning" of "harm industries" such as mining (Benson and Kirsch 2010: 461). While this criticism is abundantly warranted, Welker, Partridge, and Hardin astutely argue that "the exclusive focus on the negative aspects of these institutions undermines our ability to understand and even challenge corporate life writ large" (2011: S7). Critical analysis of extractive corporations is also stymied by condemnations that anthropomorphize these complex entities as pathological individuals who heartlessly pursue their interests without concern for anyone or anything else. This view is dramatized in legal scholar Joel Bakan's popular book *The Corporation* (2005) and associated films (Bakan and Abbott 2020; Bakan et al. 2004). An emerging body of anthropological scholarship, in contrast, critically scrutinizes corporate personhood.

Bashkow (2014: 296) clearly articulates the central paradox of corporate personhood when he writes:

> The concept of the corporate person would seem to be such a union of antithetical elements. On the one hand, the category includes some of the largest organizations on earth, great companies with multitudes of employees working in diverse roles at far-flung locations. On the other hand, it is a unity: an individual with a single identity … How does a collectivity, comprising numerous individuals, come to count as but one?

This is not a new question for Anthropology. Anthropologists since at least Maine ([1861] 1931) have pondered incorporation – that is, the creation of composite persons that are simultaneously singular and multiple – in diverse contexts. The discipline has a similarly long history of critiquing the hegemonic Euro-American conception of humans as autonomous, intentional, and self-realizing individuals. Mauss (1985) famously contrasted the individualized notion of the person in Greek, Roman, and Christian philosophy with the composite conceptions in other traditions. In response to anthropologists' continued use of corporate personhood to analyse entities organized around principles as different as kinship, politics, and territory, Appell (1983) underscored that "the corporation" is a specifically Anglo-American legal concept that should not be used cross-culturally. Kantorowicz ([1957] 1997) traces the genealogy of this form of composite personhood – from Christian theology to the English parliamentary state – that defines a corporate person as composed of two inseparable bodies: the mortal ruler and the eternal body politic. It has been the grounds for some of the mightiest entities in history, including the church, the kingdom, the empire, and the state, before being adopted in business (Shever 2012: 194–197). Although it is not universal, the corporate form is now globally dominant. The oil state of Venezuela, for example, is part of its legacy (Coronil 1997).

Many anthropologists have looked to the law to explain the power of corporate personhood in the extractive industries and beyond. The Anglo-American legal system – and the numerous other ones based on it – has granted incorporated businesses specific characteristics, including the separation of management from ownership, limited liability, and, increasingly, the same rights as human citizens (Coleman 2014; Ho 2009; Welker and Wood 2011). These features have enabled business corporations to "selectively claim the rights of natural persons while ignoring the social responsibilities of personhood" (Kirsch 2014a: 211). Chevron is one of the numerous oil companies that have deployed the corporate form to their advantage in and out of court (Sawyer 2006). There is a counter legal tradition that treats corporations as property, and therefore as objects of contractual relationships and not themselves persons (Foster 2012; Riles 2004). Yet the tradition of individualized corporate personhood has risen to prominence in the legal and financial domains, and is being extended outside them.

Corporate personhood has taken on life of its own at sites ranging from the World Bank (Merry 2011) to NGOs (Fortun and Fortun 1998) and even activist movements opposing corporate extraction. Although these movements do not usually endorse the treatment of corporations as individual people (e.g. Fitz-Henry 2018), they nonetheless engage with it and, in the process, sometimes unintentionally reinforce its power (e.g. Shever 2010). While anthropologists illustrate how corporate personhood is crucial to the hegemony of corporations, they also raise the question: *to whom* do corporations seem individual people? Golub points out that they may "appear to be unproblematically existing actors, but the closer you come to them, the more their coherence and integrity begin to falter" (2014: 12). To return to my prior example, Shell's remaking of itself into a caring neighbour and upstanding citizen was credible enough for most employees, consumers, and investors to continue to do business with the company. But it was not persuasive for Shell's "community partners" in the impoverished shanty neighbourhoods near its Argentine refinery. Shell officials nonetheless were able to deploy corporate personhood to quell opposition to the company's operations, and to deflect corporate responsibility for human welfare in the area (Shever 2010). This is but one illustration that corporate personhood is simultaneously a "fiction" in a legal sense and very real in another.

Anthropologists working in Oceania have tackled Bashkow's question of the corporation's "union of antithetical element" by drawing on Strathern's comparative theorization of personhood and exchange (e.g. Riles 2011). Kirsch (2014a) has challenged scholars to analyse corporations through a Melanesian perspective – in which people are composites of their relationships with humans and nonhuman others – rather than through a liberal European notion of self-contained individual personhood. Gordon takes on this challenge in examining how a Māori-owned corporation in New Zealand organized its business and created a 500-year business plan (2016, 385–386). Cross (2014) elaborates a Strathernian analytic with a South Asian twist to propose that corporate gifting can create corporations as persons distinct from their workers. In his example, the managers of a diamond-cutting and polishing factory in southern India understood the gold coins they gave workers to be an enticement to continue labouring for the company, while the workers viewed them as reciprocity for their prior gifts of labour. Cross concludes that corporate gifting has produced "relationships of separation and difference" as much as "forms of attachment and connection" (2014: 128). Exchange, like law, incorporates business enterprises as composite individuals while excluding many from them.

Other anthropologists have addressed the corporation's seemingly contradictory singularity and multiplicity by suggesting that they are not actually the stable and unified entities that they usually seem, despite their dominance in the extractive industries. Gordon (2016) has coined "the contingent corporation" to capture the ambiguous relationship between the legal personhood of a corporation and the people who own or represent it. Golub, likewise, has proposed conceiving of mining companies as "leviathans", powerful embodied beings that are nonetheless "potent because [they are] constructed" (2014: 13). Welker (2014) argues that corporations do not simply exist; they must be continually "enacted". Her analysis of the dynamic relationship among the US-based Newmont Mining Corporation, the Indonesian state, and community groups in Sumbawa, Indonesia, demonstrates how all three entities are continually generated in relation to one another. Her ethnography shows how Newmont is enacted through a range of practices, from construction to law and CSR. Welker thus concludes that corporations should be understood "as inherently unstable and indeterminate, multiply authored, always in flux, and comprising both material and immaterial parts" (4). My research on the Argentine oil industry has similarly led me to argue that corporations are better conceptualized as *effects of* capitalism, rather than as individual or composite actors *within* capitalism (Shever 2012: 192–199). Both Welker and I draw from Mitchell's theorization of "the state effect" to assert that corporate practices, like state ones, "create the effect of an enduring structure apparently external to those practices" (Mitchell 1999: 78). That is, the "enduring structure" of a corporation does not stand on its own. It needs to be constantly built, maintained, and refashioned through the labours of a wide range of people, both those formally within the company's boundaries and those outside them. Some corporate practices, such as the discursive and visual ones used in marketing and public relations, create a unique identity for a company: its brand persona. Other practices, such as those used in production and distribution, detach and reorganize its pieces. Oil companies that were once vertically integrated, for instance, have been divided into different subsidiaries for drilling and refining hydrocarbons, while its fuel stations and tanker transport units are converted into independently owned franchises, and its prospecting operations and oil-well maintenance are turned over to independent contractors. This enables an oil company to enlist, manage, and dismiss each one of these "dependents" to the advantage of the "parent" corporation, thus offloading its financial, environmental, and social risks. At the same time, the company's logo, branding, and advertising make this complex and ever-changing web appear to be a single entity with its own identity, values, and personality. The corporate effect thereby generates businesses that are autonomous, coherent, and bounded individuals in some respects, yet diffuse, heterogeneous, and inconstant assemblages in others (Shever 2012: 197–199). The theorization of corporations as enactments or effects explains the enduring power of extractive corporations.

6 Conclusion

Anthropology is now approaching a century of grappling with extractive corporations. The discipline has continually queried their power: from where does it come? How far does it reach? How does it work? When does it fail? Over time, anthropologists have shifted from portraying extractive corporations as outside forces that disturb the continuity of otherwise timeless societies, to recognizing them as common forces within societies the world over. They have correspondingly moved away from conceiving of mining and oil companies as personifications of the capitalist elite and towards analysing them as mutable and heterogeneous composite persons. Yet mining and oil companies are still widely assumed to be on the uppermost tier of an unwavering political-economic and social hierarchy. The CEOs of the enormous corporations that dominate extraction worldwide continue to be envisioned as sitting on gilded thrones, ruling over the world below them. This vision can be seen in Nader's appeal to anthropologists to "study up" as well as "down". Nader argued that anthropologists should investigate the small number of people in "the most powerful strata" of US society – particularly those leading companies and state agencies – because they had "the power of life and death over so many" (1974: 289 284). In the several decades since Nader's call, anthropologists have conducted extensive ethnographic research on extractive corporations. New research has led a number of anthropologists to question the dominant portrait of power and begin to redraw it. We ask what if, instead of seeing corporate executives as kings who rule from their place at the pinnacle of society, we conceive of them as precariously balanced atop an unstable structure?

Current scholarship on extraction challenges us to consider extractive corporations' power as less secure than previously thought. Anthropology now teaches that their power is not generated simply by finding, removing, and selling valuable minerals and hydrocarbons, but also by marketing them as "goods" (in the double sense of necessary commodities and social benefits), by shaping people's desires and aspirations, and by forecasting and modelling the future. Corporate power thus relies on people in their roles as consumers, taxpayers, and citizens, as well as executives, managers, workers, state agents, and shareholders. It is made and remade in public-relations campaigns and CSR projects, in negotiations and lawsuits, in subcontracting contracts, and in benefits agreements. It is concurrently circumscribed in the geographies of tax codes, environmental regulations, and physical infrastructure. In other words, corporate power is continually created through the discursive, visual, bodily, and other practices of countless people, both those situated inside extractive corporations and those formally outside them. Anthropological investigation reveals that the seemingly invincible corporations that inhabit the contemporary world are the result of contingent processes. They do not simply exist. This suggests that they can be made differently in the future. Neither the unrelenting growth of the political, economic, and social power of giant multinationals nor their continued consolidation is inevitable. Corporate power is constantly built and reinforced, but it does not have to be. Understanding corporations as potent and pervasive, but not as invincible or indestructible, provides a means of imagining extraction, and by extension global capitalism, differently.

Note

1 We should not ignore the concomitant growth of state-owned, or partially state-owned, companies, yet many of their projects involve partnerships with private corporations.

Bibliography

Appel H. (2019) *The licit life of capitalism: US oil in Equatorial Guinea*, Durham: Duke University Press.

Appell G.N. (1983) Methodological problems with the concepts of corporation, corporate social grouping, and cognatic descent group. *American Ethnologist* 10 (2): 302–311.

Apter A. (2005) *The Pan-African nation: Oil and the spectacle of culture in Nigeria*, Chicago: University of Chicago Press.

Babidge S. (2013) "Socios": The contested morality of "partnerships" in indigenous community–mining company relations, northern Chile. *The Journal of Latin American and Caribbean Anthropology* 18 (2): 274–293.

Bakan J. (2005) *The corporation: The pathological pursuit of profit and power*, New York: Free Press.

Bakan J. and Abbott J. (2020) The new corporation: The unfortunately necessary sequel. Screen Siren Pictures, Grant Street Productions.

Bakan J., Achbar M. and Abbott J. (2004) The corporation. Big Picture Media Corp.

Ballard C. and Banks G. (2003) Resource wars: The anthropology of mining. *Annual Review of Anthropology* 32 (1): 287–313.

Barrett R. and Worden D. (2014) *Oil culture*, Minneapolis: University of Minnesota Press.

Bashkow I. (2014) Afterword: What kind of a person is the corporation? *Political and Legal Anthropology Review* 37 (2): 296–307.

Bebbington A. and Bury J. (2013) *Subterranean struggles: New dynamics of mining, oil, and gas in Latin America*, Austin: University of Texas Press.

Benson P. and Kirsch S. (2010) Capitalism and the politics of resignation. *Current Anthropology* 51 (4): 459–486.

Billo E. (2015) Sovereignty and subterranean resources: An institutional ethnography of Repsol's corporate social responsibility programs in Ecuador. *Geoforum* 59: 268–277.

Carstens P. (1966) *The social structure of a Cape coloured reserve: A study of racial integration and segregation in South Africa*, Oxford: Oxford University Press.

Carstens P. (2001) *In the company of diamonds: De Beers, Kleinzee, and the control of a town*, Athens: Ohio University Press.

Chowdhury N.S. (2016) Mines and signs: Resource and political futures in Bangladesh. *Journal of the Royal Anthropological Institute* 22 (S1): 87–107.

Coleman L. (2014) Corporate identity in citizens united: Legal fictions and anthropological theory. *Political and Legal Anthropology Review* 37 (2): 308–328.

Comaroff J. and Comaroff J. (2001) *Millennial capitalism and the culture of neoliberalism,* Durham: Duke University Press.

Coronil F. (1997) *The magical state: Nature, money, and modernity in Venezuela,* Chicago: University of Chicago Press.

Coumans C. (2011) Occupying spaces created by conflict: Anthropologists, development NGOs, responsible investment, and mining. *Current Anthropology* 52 (S3): 29–43.

Cross J. (2011) Detachment as a corporate ethic: Materializing CSR in the diamond supply chain. *Focaal* 2011 (60): 34–46.

Cross J. (2014) The coming of the corporate gift. *Theory, Culture and Society* 31 (2–3): 121–145.

Davis-Floyd R. (1998) Storying corporate futures: The Shell scenarios. In: Marcus G.E. (ed.) *Corporate futures: The diffusion of the culturally sensitive corporate form.* Chicago: University of Chicago Press, 141–176.

Dinius O.J. and Vergara A. (2011) *Company towns in the Americas: Landscape, power, and working-class communities,* Athens: University of Georgia Press.

Dolan C. and Rajak D. (2016) *The anthropology of corporate social responsibility,* New York: Berghahn Books.

Evans G., Goodman J. and Lansbury N. (2002) *Moving mountains: Communities confront mining and globalisation,* London: Zed Books.

Falls S. (2014) *Clarity, cut, and culture: The many meanings of diamonds,* New York: New York University Press.

Ferguson J. (1999) *Expectations of modernity: Myths and meanings of urban life on the Zambian copperbelt,* Berkeley: University of California Press.

Ferguson J. (2005) Seeing like an oil company: Spaces, security, and global capitalism in neoliberal Africa. *American Anthropologist* 107 (3): 377–382.

Ferry E.E. (2013) *Minerals, collecting, and value across the U.S.–Mexico border,* Bloomington: Indiana University Press.

Finn J.L. (1998) *Tracing the veins: Of copper, culture, and community from Butte to Chuquicamata,* Berkeley: University of California Press.

Fitz-Henry E. (2018) Challenging corporate "personhood": Energy companies and the "rights" of non-humans. *Political and Legal Anthropology Review* 41 (S1): 85–102.

Fortun K. (2001) *Advocacy after Bhopal: Environmentalism, disaster, new global orders,* Chicago: University of Chicago Press.

Fortun M. and Fortun K. (1998) Making space, speaking truth: The Institute for Policy Studies, 1963–1995. In: Marcus G.E. (ed.) *Corporate futures: The diffusion of the culturally sensitive corporate form.* Chicago: University of Chicago Press, 249–278.

Foster R.J. (2012) Big men and business: Morality, debt and the corporation: A perspective. *Social Anthropology* 20 (4): 486–490.

Garsten C. (2012) Corporate social responsibility and cultural practices on globalizing markets. In: Kockel U., Nic Craith M. and Frykman J. (eds) *A companion to the anthropology of Europe.* Hoboken: Wiley-Blackwell, 407–424.

Godoy R. (1985) Mining: Anthropological perspectives. *Annual Review of Anthropology* 14 (1): 199–217.

Golub A. (2014) *Leviathans at the gold mine: Creating indigenous and corporate actors in Papua New Guinea,* Durham: Duke University Press.

Gordon G. (2016) Culture in corporate law or: A black corporation, a Christian corporation, and a Maori corporation walk into a bar. *Seattle University Law Review* 39 (2): 353–396.

Gusterson H. (1997) Studying up revisited. *Political and Legal Anthropology Review* 20 (1): 114–119.

Guyer J.I. (2011) Blueprints, judgment, and perseverance in a corporate context. *Current Anthropology* 52 (S3): 17–27.

Harvey D. (2005) *A brief history of neoliberalism*, Oxford: Oxford University Press.

High M.M. (2019) Projects of devotion: Energy exploration and moral ambition in the cosmoeconomy of oil and gas in the western United States. *Journal of the Royal Anthropological Institute* 25: 29–46.

Ho K. (2009) *Liquidated: An ethnography of Wall Street*, Durham: Duke University Press.

Huber M.T. (2013) *Lifeblood*, Minneapolis: University of Minnesota Press.

Jacka J.K. (2015) *Alchemy in the rain forest: Politics, ecology, and resilience in a New Guinea mining area*, Durham: Duke University Press.

Jacka J.K. (2018) The anthropology of mining: The social and environmental impacts of resource extraction in the mineral age. *Annual Review of Anthropology* 47 (1): 61–77.

Kantorowicz E.H. ([1957] 1997) *The king's two bodies: A study in mediaeval political theology*, Princeton: Princeton University Press.

Kirsch S. (2014a) Imagining corporate personhood. *Political and Legal Anthropology Review* 37 (2): 207–217.

Kirsch S. (2014b) *Mining capitalism: The relationship between corporations and their critics*, Berkeley: University of California Press.

Kirsch S. (2020) Between the devil and the deep blue sea: Objectivity and political responsibility in the litigation of the Exxon Valdez oil spill. *Critique of Anthropology* 40 (4): 403–419.

Li F. (2015) *Unearthing conflict: Corporate mining, activism, and expertise in Peru*, Durham: Duke University Press.

Maine S.H. ([1861] 1931) *Ancient law*. London: Dutton.

Marcus G.E. (1995) Ethnography in/of the world system: The emergence of multi-sited ethnography. *Annual Review of Anthropology* 24: 95–117.

Mason A. (2013) Cartel consciousness and horizontal integration in energy industry. In: Strauss S., Rupp S. and Love T.F. (eds) *Cultures of energy: Power, practices, technologies*. Walnut Creek: Left Coast Press, 126–138.

Mauss M. (1985) A category of the human mind: The notion of person; the notion of self. In: Carrithers M., Collins S. and Lukes S. (eds) *The category of the person: Anthropology, philosophy, history*. Cambridge: Cambridge University Press, 1–25.

Merry S.E. (2011) Measuring the world: Indicators, human rights, and global governance. *Current Anthropology* 52 (S3): 83–95.

Mitchell T. (1999) Society, economy and the state effect. In: Steinmetz G. (ed.) *State/culture: State formation after the cultural turn*. Ithaca: Cornell University Press, 76–97.

Muñoz J-M. and Burnham P. (2016) Subcontracting as corporate social responsibility in the chad-cameroon pipeline project. In: Dolan C. and Rajak D. (eds) *The anthropology of corporate social responsibility*. New York: Berghahn Books, 152–178.

Nader L. (1974) Up the anthropologist: Perspectives gained from studying up. In: Hymes D. (ed.) *Reinventing anthropology*. New York: Vintage Books, 284–311.

Nash J.C. (1979) Anthropology of the multinational corporation. In: Gerrit H. and Bruce M. (eds) *The politics of anthropology: From colonialism and sexism toward a view from below*. The Hague: Mouton, 421–446.

Nash J.C. ([1979] 1993) *We eat the mines and the mines eat us: Dependency and exploitation in Bolivian tin mines*, New York: Columbia University Press.

Porteous J.D. (1974) Social class in Atacama company towns. *Annals of the Association of American Geographers* 64 (3): 409–417.

Proctor R.N. (2001) Anti-agate: The great diamond hoax and the semiprecious stone scam. *Configurations: A Journal of Literature, Science, and Technology* 9 (3): 381–412.

Rajak D. (2008) "Uplift and empower": The market, morality and social responsibility on South Africa's platinum belt. In: Luetchford P., Pratt J., Wood D.C., et al. (eds) *Hidden hands in the market: Ethnographies of fair trade, ethical consumption and corporate social responsibility*. Bingley: Emerald, 297–324.

Rajak D. (2011) *In good company: An anatomy of corporate social responsibility*, Stanford: Stanford University Press.

Rajak D. (2014) Corporate memory: Historical revisionism, legitimation and the invention of tradition in a multinational mining company. *Political and Legal Anthropology Review* 37 (2): 259–280.

Rajak D. (2016) Expectations of paternalism: Welfare, corporate responsibility, and HIV at South Africa's mines. *South Atlantic Quarterly* 115 (1): 33–59.

Richards A.I. ([1939] 1995) *Land, labour and diet in northern Rhodesia: An economic study of the Bemba tribe*, Münster: LIT Verlag.

Riles A. (2004) Property as legal knowledge: Means and ends. *Journal of the Royal Anthropological Institute* 10 (4): 775–795.

Riles A. (2011) Too big to fail. In: Edwards J. and Petrović-Šteger M. (eds) *Recasting anthropological knowledge: Inspiration and social science*. Cambridge: Cambridge University Press, 31–48.

Rogers D. (2015) *The depths of Russia: Oil, power, and culture after socialism*, Ithaca: Cornell University Press.

Rubbers B. (2015) When women support the patriarchal family: The dynamics of marriage in a Gécamines mining camp. *Journal of Historical Sociology* 28 (2): 213–234.

Ruggie J.G. (2002) The theory and practice of learning networks. *The Journal of Corporate Citizenship* (5): 27–36.

Sawyer S. (2001) Fictions of sovereignty: Of prosthetic petro-capitalism, neoliberal states, and phantom-like citizens in Ecuador. *Journal of Latin American Anthropology* 6 (1): 156–197.

Sawyer S. (2004) *Crude chronicles: Indigenous politics, multinational oil, and neoliberalism in Ecuador*, Durham: Duke University Press.

Sawyer S. (2006) Disabling corporate sovereignty in a transnational lawsuit. *Political and Legal Anthropology Review* 29 (1): 23–43.

Sawyer S. (2010) Human energy. *Dialectical Anthropology* 34 (1): 67–75.

Shever E. (2008) Neoliberal associations: Property, company and family in the Argentine oil fields. *American Ethnologist* 35 (4): 701–716.

Shever E. (2010) Engendering the company: Corporate personhood and the "face" of an oil company in metropolitan Buenos Aires. *Political and Legal Anthropology Review* 33 (1): 26–46.

Shever E. (2012) *Resources for reform: Oil and neoliberalism in Argentina*, Stanford: Stanford University Press.

Shever E. (2013) "I am a petroleum product": Kinship, sentiment and capitalism in the Argentine oil industry. In: McKinnon S. and Cannell F. (eds) *Vital relations: Modernity and the persistent life of kinship*. Santa Fe: School for Advanced Research Press, 85–107.

Sydow J. (2016) Global concepts in local contexts: CSR as "anti-politics machine" in the extractive sector in Ghana and Peru. In: Dolan C. and Rajak D. (eds) *The Anthropology of Corporate Social Responsibility*. New York: Berghahn Books, 217–242.

Tinker Salas M. (2009) *The enduring legacy: Oil, culture, and society in Venezuela*, Durham: Duke University Press.

Urban G. and Koh K-N. (2012) Ethnographic research on modern business corporations. *Annual Review of Anthropology* 42: 139–158.

Vitalis R. (2007) *America's kingdom: Mythmaking on the Saudi oil frontier*, Stanford: Stanford University Press.

Watts M. (2005) Righteous oil? Human rights, the oil complex and corporate social responsibility. *Annual Review of Environment and Resources* 30 (9): 373–407.

Welker M. (2014) *Enacting the corporation: An American mining firm in post-authoritarian Indonesia*, Berkeley: University of California Press.

Welker M., Partridge D.J. and Hardin R. (2011) Corporate lives: New perspectives on the social life of the corporate form. *Current Anthropology* 52 (S3): 3–16.

Welker M. and Wood D. (2011) Shareholder activism and alienation. *Current Anthropology* 52 (S3): 57–69.

Weszkalnys G. (2015) Geology, potentiality, speculation: On the indeterminacy of first oil. *Cultural Anthropology* 30 (4): 611–639.

Yessenova S. (2012) The Tengiz oil enclave: Labor, business, and the state. *Political and Legal Anthropology Review* 35 (1): 94–114.

Zalik A. (2004) The peace of the graveyard: The voluntary principles on security and human rights in the Niger Delta. In: van der Pijl K., Assassi L. and Wigan D. (eds) *Global regulation: Managing crises after the imperial turn*. Basingstoke: Palgrave Macmillan, 111–127.

3

DEVELOPMENT

Robert Jan Pijpers

1 Introduction

Resource extraction is generally considered fundamental to processes of economic growth, and a key element in trajectories out of poverty and towards prosperity. Examples are abundant. In 20th-century Venezuela, for instance, oil extraction's potential generated visions that "instantaneous modernization lay at hand" (Coronil 1997: 10), while in Australia, mining is central to ideas of development and nation building as it contributes to the creation of "a civilized (and hence morally commendable) social life in the context of economic growth and progress" (Trigger 1997: 164). Correspondingly, in contemporary Poland politicians reassert coal mining "as a bedrock of national development" (Kuchler and Bridge 2018: 136), in Greenland there has recently been a political turn towards viewing the country's mineral sector as the basis of long-term development and economic and political independence (Nuttall 2013), and in Sierra Leone large-scale mining has been positioned as a key element in the country's post-war development agendas and trajectory of progress (Pijpers 2016a). As evident from these examples, the processes of growth and progress that resource extraction is projected to generate are frequently framed in terms of development.

Although not entirely new, the idea of development gained traction after WWII and "went on to become one of the dominant ideas of the 20th century" (Lewis 2005: 472). This period – often framed as the "era of development" (Thomas 2000: 5) – was characterized by a strong belief in growth and progress and a conviction that all societies were heading towards a modern condition modelled on the West (Arce and Long 2000: 5–6). The work of economist Rostow (1960 [1990]), which delineates five stages of economic growth through which societies must pass in order to become modern, is exemplary in terms of this idea of modernization. The model, which has been highly influential in development

DOI: 10.4324/9781003018018-3

thinking,[1] sees natural resource wealth and its extraction as a key condition for industrial "take-off", sparking development and putting societies on a trajectory towards modernity.

While such ideas of growth typically dominate common understandings of development, the concept has simultaneously acquired several further layers of meaning. In synthesizing these, Lewis (2005: 474) distinguishes four: 1) activities required to bring about change or progress, often linked to economic growth; 2) a standard against which different rates of progress may be compared, thereby taking on a subjective, judgemental element; 3) planned social change (with its objectives, I would like to add, articulated in specific frameworks, such as the Sustainable Development Goals)[2] and the idea of external intervention; and 4) within radical critiques, an organized system of power and practice which has formed part of colonial and neo-colonial domination (Lewis 2005: 474; see also Allen and Thomas 2000; Edelman and Haugerud 2005; Escobar 1997; Ferguson 1990; Mosse 2013). Of course, these meanings are not disconnected; rather, they overlap and offer different vantage points from which to approach the concept of development and its associated visions, ideologies, power relations, mechanisms, and effects. While staying attuned to this diversity, in the specific context of this chapter we focus on development's meaning as an anticipated process of growth and progress enabled by the extraction of natural resources.

While resource extraction has long been part of processes of growth, including in the eras of imperialism and colonialism (Edelman and Haugerud 2005; Jacka 2018), with the emergence of the contemporary rhetoric of development it has also been increasingly framed in these terms. In addition to the examples mentioned earlier, this transpires in the post-independence development trajectories of countries such as India (Lahiri-Dutt 2016) and Ghana (Hilson 2002), and extraction's prominence in Structural Adjustment Programmes (SAPs): a set of neoliberal policies conditioning loans which were rolled out in the 1980s and 1990s in developing countries under the aegis of the World Bank (WB) and the International Monetary Fund (IMF). In contrast to the earlier development approaches of the WB and IMF, which had encouraged state interventions in the economy (Edelman and Haugerud 2005: 6–7), the SAPs sought to reduce the role of the state by promoting forms of market liberalization, deregulation, and privatization. These reforms significantly affected the (regulation of) extractive industries and (re) opened their doors to foreign investors (Campbell 2009; Fraser and Larmer 2010; Hilson 2004; Sawyer and Gomez 2012). After the turn of the millennium, the importance of resource extraction as a catalyst for development continued to be stressed by both multinational institutions and industry organizations (e.g. Cavalcanti et al. 2016; ICMM 2016; Union 2009; Weber-Fahr et al. 2001). Despite strong critiques on, among other things, the power inequalities associated with development and the growth-based ideology underpinning it, such as articulated in post-development theories (e.g. Escobar 2015; Rahnema and Bawtree 1997 [2006]), it also continues to surface in national development agendas, whether these continue to follow the SAPs neoliberal line in the form of Poverty

Reduction Strategy Papers, as in Sierra Leone's Agenda for Prosperity (GoSL 2013), attempt to reinsert state control in the extractive industries through forms of resource nationalism such as in Indonesia (Warburton 2017) or by adopting neoextractivist models, as in several Latin American countries (Gudynas 2010), or situate resource extraction as key element in the pursuit of economic and political independence, as is the case in Greenland (Nuttall 2013) or New Caledonia (Horowitz 2004). Notwithstanding the diversity among such development agendas, not least in terms of resource ownership and control, the extraction of natural resources is seen as important to processes of growth and progress in all of them. Indeed, as underscored by Gamu et al.'s (2015) critical analysis of the linkages between the two, extraction is envisioned to reduce poverty, trigger growth, and, ultimately, stimulate development through increased (fiscal) revenue flows to (central) governments, employment opportunities, investments in public sectors and goods, the stimulation of (local) economies through upstream/downstream linkages and trickle down effects, and local development initiatives.

As extraction is often framed in terms of development, anthropologists researching it frequently find themselves studying how extraction-led development is anticipated, experienced, and negotiated by different actors, and eventually materializes (or not) in specific places.[3] Typically, they have analysed its ambiguities and negative effects, critically examining dynamics related to resource curse phenomena (see Gilberthorpe and Rajak 2017). These dynamics stand in sharp contrast to the expectations of progress and prosperity that are not only part of national imaginaries and development policies, but also prominent in the spaces where extraction takes place or is projected to do so. These two aspects of extraction-led development are discussed in the first two sections of this chapter. The third section focuses on relatively recent initiatives designed to overcome the negative effects of extraction and increase development opportunities (part of what Banks et al. 2013, following the work of Cowen and Shenton 1996, would call forms of intentional development). Together, these three sections bring home the complex (local) social, economic, and political dynamics, including issues related to access, control, and redistribution, that underpin and emanate from extraction-led development.

2 Promises and expectations

While national imaginaries and development blueprints promote extraction as an opportunity for achieving development, its promises of progress, modernity, and the eradication of poverty also reverberate in people's expectations in the places where extraction eventually materializes, or is projected to do so. In Bangladesh, for example, inhabitants of villages surrounding a Chevron gas operation welcomed extraction as it was assumed it would bring jobs and promised a "desirable landscape of 'development', consisting of reliable infrastructure, brick houses, a dependable power supply, and industrialisation" (Gardner et al. 2014: 9). Similarly, in Burkina Faso, Luning (2012) shows that interactions between staff of a gold exploration company and local residents consisted specifically of "assessing the gap between what is and what may become" (ibid.: 24). Moreover, notwithstanding their concerns, local authorities viewed "favorably the presence of the mining company, as this could contribute to the development of the village [through] obtaining paid jobs and water" (ibid.: 29). In the case of Bolivia, lithium mining has generated high expectations in areas located close to (projected) mine sites, with residents anticipating a regional industrial boom leading to "beautiful streets and no garbage" (Revette 2017: 157). Crucially, Revette argues, despite Bolivia's rather negative history of extraction, the government's attribution of the failure of development to the control exercised by foreign companies, and its recasting of extraction in a discourse of resource nationalism, has led to the recalibration of hope and generated expectations among citizens that "this time it's different".

Of course, these observations regarding expectations do not imply that there is harmonious alignment among groups anticipating extraction. On the contrary, they typically vary greatly. Studying extraction-led development in Mozambique, Wiegink (2018), for example, shows that industry professionals, local business-people and members of resettled communities all have diverging ideas about the kind of development coal mining should bring. In the case of iron ore mining in Sierra Leone, Pijpers (2020) highlights contradictions between the views of corporate staff, local community members, and political authorities of various kinds, discussing how the area's past serves as a political instrument to legitimize expectations and claim desired futures. Camacho (2016) draws attention to intergenerational differences in mining communities in Chile, showing the diverging positions of the aged, the middle-aged, and the younger generations. Meanwhile, in northern Fennoscandia, Komu (2019) juxtaposes the critical positions towards extraction-led development of the tourism sector with those of pro-resource development groups, who (inspired by the region's history of extraction and related developmental milestones) expect that mining "would boost the local economy, increase the population, and raise the general spirits in the municipality" (Komu 2019: 122).

In addition to large-scale extraction contexts such as those discussed above, expectations of prosperity are equally prominent in the artisanal and small-scale mining (ASM) sector, although here they are typically articulated at a more individualistic level. Illustratively, and in resonance with other work (D'Angelo 2019; Geenen 2018; Jaramillo 2020; Pijpers 2017), artisanal gold and diamond miners in Tanzania perceive mining as "an avenue for the pursuit of a better life" and "a ladder that sends us wealth" (Fisher et al. 2009). And indeed, although often overlooked in development agendas (Hilson and McQuilken 2014) – hence the focus on large-scale mining in this chapter – ASM provides income-earning opportunities for at least 100 million people globally (Pelon and Walser 2009) and, with formalization often considered a prerequisite (Hilson et al. 2021), has the potential to contribute significantly to poverty reduction (Fisher et al. 2009)

As indicated, for example by Trigger's (1997) work on extraction-led development in Australia (mentioned above), it is key to stress that expectations do not only involve material, but also social progress (though the two are often connected). An excellent example is provided by Ferguson's *Expectations of Modernity* (1999), discussed in more detail in Section 3, which demonstrates that Zambians expected their country to move proudly "into the ranks of the first class" (ibid.: 13) and become part of the new world society (ibid.: 235) following the copper mining boom in the early and mid-20th century. Correspondingly, Schreer's work on gold mining in Indonesian Borneo shows how gold mining offers opportunities for people to "take up their place as legitimate participants in the possibilities arising from modernity as well as its ongoing inequalities of power, resources, and opportunities by establishing convivial relations with others to reproduce themselves as modern citizens" (Schreer 2020: 3). Other work addresses anticipations that extraction will offer opportunities for reconfiguring constraining relationships and structures and, as a result, will contribute to forms of social progress. This view is common, for example, among men in Tanzania who aspire to live free of the oppressive relations that characterize life in their home towns (Jønsson and Bryceson 2014), among women in Colombia who, aside from the problem of gender-based violence, find empowerment in the artisanal gold mining sector (Cohen 2014: 272), and among women sapphire miners in Madagascar who often migrate to mining sites to "look for a new life in a society where gender inequalities are smaller than in the other parts of the country" (Canavesio 2013: 1).

In sum, anthropologists focusing on the promises and expectations of extraction-led development have brought home that these are not only central to policies and national imaginaries, but also prominent in the places where extraction-led development materializes. Moreover, scholarship illuminates how expectations inform a range of social dynamics and anticipatory actions. These include negotiations between different actors and the (attempted) reconfiguration of gender relations and ways of participating in (global) society, as illustrated in several works discussed above, but also by the establishment of complete apparatuses of expectation-management, as is the case in São Tomé e Príncipe (Weszkalnys 2008); by domestic and transnational mobility towards extraction sites, whether in Mongolia (High 2017), Tanzania (Bryceson and Fisher 2014), the Guianas (de Theije 2016) or Melanesia (Koczberski and Curry 2004); and by reconfigurations in the domains of land ownership and governance practices like those in Papua New Guinea (Jacka 2016). Consequently, interrogating the expectations of extraction-led development is key, as it contributes to understandings of the processes and effects triggered by extractive projects and their development promises.

3 Problems and ambiguities

While expectations of growth and progress characterize the visions and vistas of extraction-led development, probably the largest share of anthropological work in the extraction–development nexus is dedicated to its wide range of ambiguities. Building on broader critiques of modernization and development – which centralize issues of power and politics in relation to inequality and development (e.g.

Escobar 1988; Ferguson 2006; Frank 1967) – these ambiguities are typically approached as (local) manifestations of the resource curse (Auty 1993). However, this is done by understanding the resource curse in its broadest sense – that is, as "the paradoxical situation where what should bring good brings bad" (Reyna and Behrends 2008: 11) – and, crucially, in ways that go beyond technical questions of revenue management but rather foreground relations of power and remain critical towards the "assumptive basis and unilinear teleology" of the notion of the resource curse (Gilberthorpe and Rajak 2017: 189). The manifestations of extraction's negative impacts and ambiguities come in different forms and play out at different scales, as Bainton (2020) discusses, ranging from violent conflict, social upheaval, and exclusion from decision-making processes, to grievance, community conflict, loss of livelihoods, environmental damage, and the severe effects (on health) of pollution. To avoid turning this section into a long list of problems and ambiguities, I draw out four domains in which the limitations of extraction's development potential have been addressed. Given that the promises of extraction-led development are typically articulated in terms of increased resource flows, these four domains – (trans)national financial flows, employment, land, and extraction's aftermath – are all tied to redistributive practices and draw attention to the issue of where and to whom resources flow.

A major assumption of extraction-led development is that extractive activities stimulate the development of local economies through upward/downward linkages and trickle-down effects. Yet numerous scholars have problematized these theories, not least by foregrounding the enclaved nature of extraction (Akiwumi 2012; Appel 2012; Yessenova 2012). Exemplary is Ferguson's (2005) analysis of the territorialization of capital investment (oil serves as his example) in secured enclaves with little or no economic benefit to the wider society. To support this argument, Ferguson demonstrates how capital "does not 'flow' from London to Cabinda [but] hops, neatly skipping over most of what lies in between" (ibid.: 379). By foregrounding the division between pockets of investment and their wider surroundings, and the efforts that corporations make "toward the abdication of responsibility for life 'outside' their walls" (Appel 2012: 460), enclave theories challenge trickle-down principles, showing that extraction-led growth is contingent on global power structures that create forms of local separation rather than integration.

Aside from transnational resource flows (or hops), scholars have also focused on the resource flows at national and local levels and the politics of control directing them. Stammler and Wilson (2006: 15–16), for example, discuss how the Russian government has re-exerted control over the energy sector by centralizing oil taxation and redistributing oil revenues nationally. Consequently, decentralized governing bodies have seen their budgets drying up and fear that local expectations of improved living standards will not materialize. Other scholars have drawn attention to processes of rent-seeking, patronage, and clientelism, which can take shape at the national level due to close connections between corporations and powerful individuals within the state apparatus (for a Malaysian and Russian example see Gomez 2003, and Rogers 2014, respectively), or at the local level where elites can exercise informal control over mining operations and the revenues they generate. The effects of these processes can be multiple and ambiguous. For example, the rent-seeking behaviour of local strongmen in small-scale mining in eastern Mindanao in the Philippines leads to private gains, but also to a certain regulation of the sector (Verbrugge 2015). Additionally, while individual financial gains may consolidate the power of local elites, as Welker (2009) has shown in the case of Indonesia, it may also weaken their positions as they are subjected to unrealizable expectations to share their wealth (Bainton 2010) as well as greater criticism (Ahasan and Gardner 2016). Yet, whatever the case, the part played by elites in controlling resources generates significant opportunities for some, but also poses challenges to those who lack the necessary connections.

The role of elites is also important in the domain of employment, which is among the primary promises of extraction-led development, as they can direct jobs (and other opportunities such as contracts for transport or catering services) towards themselves and their acquaintances (Welker 2014). In the case of Papua New Guinea, for example, extractive companies employ well-connected and powerful individuals in executive positions to gain traction in the bureaucratic apparatus of the state, to reduce potential frictions, and to "ward off irregularities and contingencies that might draw the company down" (Golub and Rhee 2013: 232). In addition to these phenomena, the dynamics generated by the temporality of extractive operations, such as those related to technological shifts, influence where and to whom this particular resource flows and, consequently, who stands to benefit. As studies from Bangladesh (Ahasan and Gardner 2016), Sierra Leone (Pijpers 2020), and Russia (Stammler and Wilson 2006) have shown, extractive operations typically require a substantial number of low-skilled labourers in the early phases. However, as operations move into production or shift to different technologies, labour demands can decrease, resulting in lay-offs. Moreover, the advancement of projects can also involve a shift in the composition of the labour force, as "locals", for example, may be replaced by domestic or international "strangers" with higher education levels (the idea of fly-in, fly-out workers is telling in this regard). Furthermore, such dynamics have an important gendered dimension, as Ahmad and Lahiri-Dutt (2006: 331) show in India where mechanization developments have pushed women out of the mining industry (see also D'Angelo, this volume, for more detailed discussion). Such dynamics may result in disillusionment among local community members over the limited availability of jobs, which are often temporary and low paid, as well as giving rise to local tensions between those who have jobs and those who do not and "who may be struggling to maintain their prior standard of living in the face of rising costs of housing and goods" (Stammler and Wilson 2006: 16–17). Consequently, while employment opportunities are a key element of the promise of extraction-led development, it is crucial to ask for whom these opportunities materialize, at what stage of extraction, and with what social effects.

Another vital aspect of the effects/limitations of extraction-led development centres around issues related to land. Here again, local power divisions, often connected to forms of belonging and positions of landowners versus land users, affect the redistribution of benefits and costs and can lead to tense social processes. Exemplary in this regard is Jacka's (2018) discussion of landed power dynamics in Papua New Guinea. As customary landowners are entitled to share in the benefits of extractive activities on their land, being able to claim to belong to landowning families or clans yields great benefits, while not being able to do so leads to exclusion from certain entitlements. As a result, other strategies of claiming benefits emerge. At Papua New Guinea's Porgera mine, for example, young men worked "in the lifemarket", which meant "engaging in interclan warfare with groups who had received money from the mining project to try and extort money from them, or at least die trying, knowing that their lives would then have to be compensated by the group they attacked" (Jacka 2018: 70, see this reference for details of the work of Banks 2008, Jorgensen 2004, Jacka 2015, and Macintyre and Foale 2007 which inform this analysis). In other words, land ownership structures inform revenue flows which, in turn, trigger fierce negotiations over belonging and landed power, and, in the slipstream, social tensions between those who can secure (legal) entitlement and those who cannot.

Landed (power) relations, and landed dynamics more broadly, are also prominent in the issue of mining-induced displacement (see, for example, Owen and Kemp 2015). Although the land targeted for extraction is sometimes portrayed as empty and vacant in order to legitimize resource extraction (Leonard 2016; Voyles 2015), in other cases land needs to be rendered vacant and accessible, leading to forced resettlement of complete communities. This has significant economic and social effects, such as the emergence of new inequalities and the disruption of livelihoods, social bonds, and the cultural roots of communities (Ahmad and Lahiri-Dutt 2006; Teaiwa 2014). It may also disproportionately affect women, as they tend to rely on resources available in their immediate surroundings while, simultaneously, (financial) compensation typically targets men due to women's exclusion from land ownership (Ahmad and Lahiri-Dutt 2006: 315). Moreover, besides being an unintended negative consequence of development, development discourses can also emerge to justify operations and enable dispossession to take place, as Ahasan and Gardner (2016) have argued. Consequently, development dispossession, or "dispossession by development" does not necessarily only refer "to the side effect of people losing land and livelihoods", but also "to the use of discourses of development ... to enable dispossession to take place" (ibid.: 13).

Finally, a number of scholars have studied the ambiguous aftermath of extraction (Chaloping March 2017; Bainton and Jackson 2020; Rubbers 2017). These studies typically show that development achievements often turn out to be unsustainable. Most prominent in this regard is Ferguson's (1999) analysis of the modernization myth and the experience of decline in the Zambian Copperbelt (once the locus of Manchester School/Rhodes Livingstone Institute studies on social change). Ferguson shows how, following the economic crisis of the 1980s, the promises and manifestations of modernity (eating meat, owning a car, electrification), which were within reach for ordinary workers, "had been abruptly yanked away" (1999:

13). Zambians felt that "the promises of modernization had been betrayed, and that they were being thrown out of the circle of full humanity" (ibid.: 236), an experience Ferguson analyses as "abjection" to stress the idea of expulsion rather than exclusion and of loss rather than lack (ibid.: 237–238). Consequently, along with other scholarship, Ferguson's analysis brings home the instability of extraction-led development, as the closure of extractive activities can generate development busts in which earlier achievements and expectations are shattered.

In sum, as the discussion above shows, extraction-led development is marked by high degrees of ambiguity and contestation, both when extraction materializes and also when it disappears. Effects and outcomes, including the potential creation of new inequalities or the exacerbation of existing ones, are contingent in complex processes guiding the redistribution of costs and benefits, whether this applies to the formation of enclaves, elite control over resources like money and jobs, local land ownership and governance structures, displacement and dispossession, or the experience of both exclusion and abjection. Thus, following Gilberthorpe and Rajak's (2017) profound analysis of critical anthropological perspectives on the resource curse, anthropological work reminds us that, although the ultimate outcomes of extraction-led development are disparate and in flux, we must stay attuned to the social relations, negotiation processes, and deeply unequal power relations that influence *how* opportunities and problems are redistributed and *how* development materializes.

4 Bridging the discrepancy, guiding the development trajectory

As a result of growing awareness of extraction's adverse impacts and limited positive effects, there has been, largely since the turn of the millennium, a proliferation of initiatives geared towards making extraction less harmful and raising its contributions to development. Ideally, this should be "sustainable development", the kind "that meets the needs of the present without compromising the ability of future generations to meet their own needs" (WCED 1987; commonly known as 'the Brundtland report', see Lanzano this volume). In the case of extraction, however, this roughly translates into the promotion of "the industry's actors as agents of 'best practice' in economic, social, environmental and governance spheres" (Gilberthorpe et al. 2016: 2).

The responses to the challenges of overcoming extraction's problems and delivering on its promises are multiple and include initiatives such as regaining control over natural resources by national states; enhancing transparency and good governance (for example the Extractive Industries Transparency Initiative); seeking and obtaining local communities' free prior and informed consent; reducing the risk that mineral revenues finance conflict (such as the Kimberly Process Certification Scheme); making supply chains more fair (fair trade and fair mined gold); taking Corporate Social Responsibility initiatives; implementing community development programmes; performing impact assessments; initiating public–private partnerships; and promoting better social, economic, and environmental practices. Telling for the latter initiative are the 59 international standards/tools on responsible mining that have, with the exception of three, been introduced since 2000 (IFC 2014: 47–54). In studying these initiatives, anthropologists have focused on their politics and effects, thereby foregrounding the social dynamics in which they are contingent, and re-politicizing what is, like other development discourse and projects, often rendered apolitical (Ferguson 1990, although see Mosse 2005: 2–6, for a critical response). In what follows, I illustrate this by focusing on: 1) consultation processes and free prior and informed consent; 2) CSR initiatives; and 3) the involvement of civil society in extractive projects.

Free prior and informed consent (FPIC) can be seen as a form of engagement and decision making in which "the free and informed consent of an indigenous people is obtained in advance of a course of action" (Szablowski 2010: 111). As several scholars have discussed (e.g. Owen and Kemp 2014; Sawyer and Gomez 2012; Szablowski 2010), the principle has been driven by the global indigenous rights movement to address the exclusion of both the institutions and interests of indigenous people and to advance their autonomy and self-determination. It has also been adopted by a wide set of regulatory frameworks of governments, corporations, and multilateral institutions, including some in the extractive industries (which are strongly associated with indigenous peoples' rights abuse; see also Nuttall this volume).

Yet, while designed to increase indigenous communities' participation in and influence on development trajectories, various scholars have critically interrogated the idea of FPIC, particularly its typical limitation to consultation rather than consent (Szablowski 2010: 117–121), and the degree of influence that can actually be exercised. For example, by drawing attention to a wide range of constraining issues and practices in the context of two mining-related consultation meetings in Bolivia, Perreault (2015) analyses consultation processes as a form of political performance. Rather than promoting meaningful participation in development and its associated decision-making processes, Perreault argues, these processes seemed designed to provide "the appearance of participatory governance, while permitting extractive activities to continue apace" (ibid.: 447). In a similar vein, Schilling-Vacaflor and Eichler (2017) discuss the "shady side" of consultation, FPIC, and compensation processes in Bolivia. As a result of divide and rule tactics applied by states and corporations, these procedures tend to benefit only a few individuals and can easily lead to the exacerbation of conflict, the weakening of indigenous organizational structures, and the division of communities. In addition, Macintyre's analysis from Papua New Guinea (Macintyre 2007) brings home that, aside from states and corporations, different local actors can also orchestrate consultation in FPIC processes and draw on internal disputes and power inequalities in ways that advance their interests. Such dynamics, Macintyre makes explicit, render questions like "who is to give consent?" and "who is the community?" (and, ultimately, "who decides about the implementation and direction of extraction-led development?") nearly impossible to answer.

Questions of politics, practices, and effects that are key to FPIC and consultation mechanisms, are also pertinent to debates about Corporate Social Responsibility

(CSR), which has "become established as orthodoxy within the arena of both development and multinational business" (Dolan and Rajak 2016: 3; see also Shever and Lanzano this volume). Although rooted in longer histories of corporate responsibility, CSR in its current form emerged in the late 1990s and can be seen as a set of practices and discourses through which corporations position themselves as "ethical actors" and (arguably) contribute to local development (Dolan and Rajak 2016: 5). In practice, this involves a variety of initiatives, ranging from constructing infrastructure and providing scholarships to neutralizing environmental damage and contributing to educational programmes.

Anthropology has approached CSR from two main vantage points: its apparatus and architecture, and its local effects, contestations, and responses (Dolan and Rajak 2016: 2). Illustratively, scholars have studied corporate incentives to implement CSR (Welker 2009; Banks et al. 2013), the weaknesses and ownership structures of CSR practices (Gilberthorpe and Banks 2012), the social and ethical dynamics of CSR's principle of transparency (Babidge 2015), and how "the gift of CSR" ties into relations of patronage, power, and dependency (Rajak 2006; see also Cross 2014). Some scholars, such as Sharp (2006), have analysed CSR as an outsourcing of development responsibilities from the state to corporations, and asked about the implications of this shift in responsibility. In pursuing this debate, Sharp shows that CSR discourse differs from earlier development discourse in the terms it uses to construct people's entitlement to development: from entitlement based on citizenship under state-led development to a categorization in which entitlement is transferred to those who belong to the corporation's host communities (ibid.: 216). Not only does the responsibility of corporations towards those who belong inspire new forms of corporate patronage and its associated inequalities (Rajak 2006: 199), it also (re)inspires local identity politics evolving around ideas of belonging and autochthony (cf. Geschiere and Nyamnjoh 2000), and gives rise to a "politics of localness" (Pijpers 2016b). The latter exposes the myriad ways through which different groups and individuals try to claim positions as being local (as part of those who belong) in order to gain access to the opportunities of corporate-led development.

The proliferation of new development tools in the extractives industries also gives rise to new partnerships between corporations and civil society organizations and/or local community groups (who are otherwise typically associated with resistance, see also Li and Velásquez this volume). Anthropologists have approached these new public–private partnerships from various angles. In the case of environmentalist groups, for example, Kirsch highlights how corporations seek out collaboration with the "light greens who view the market as the solution to environmental problems" instead of the dark greens – a strategy described as "the corporate version of divide and conquer" (Kirsch 2010: 296). Additionally, Rajak shows how, in South Africa, NGOs compete with each other to become "the partner of choice" for platinum mining corporations, a competition which weakens them "as it strips them of the autonomy under the banner of empowerment" (Rajak 2011: 192). Correspondingly, in Bangladesh, corporate partnerships not only allowed NGOs to access new sources of funding and carry out their work, but also placed them in a compromising role as actors working towards a corporate agenda, while *their* staff, rather than Chevron's, became positioned at the frontline of complaints from local people (Ahasan and Gardner 2016). In contrast, Smith Rolston (2015) brings to the fore a case of "critical collaboration" between a grassroots environmental coalition and a gold mining company in North America in which community groups maintain their autonomy. However, while this collaboration style is promising, caution is required, Smith Rolston stresses, not least because the relationship with the mine threatened the community organization's group identity, while such collaborations and partnerships risk erasing "histories of opposition and dispossession" (Sawyer 2004: 4).

In sum, anthropological analyses of the new initiatives for development proliferating since the turn of the millennium have largely focused on their politics and local effects. While there is debate about whether such initiatives are forms of corporate greenwashing or real development intentions (Jacka 2018: 65), there is certainly broad agreement on the high levels of ambiguity surrounding them, their political mobilization by those designing and those seeking to benefit from them, and the necessity to study how their operationalization and implementation pose new challenges to, and introduce new constraints on, promoting development. Consequently, questions such as "development on whose terms?" (cf. Gilberthorpe and Banks 2012) and "development for whom?" continue to be absolutely critical.

5 Conclusion

This chapter has focused on three key aspects of extraction-led development: 1) the promises and expectations of progress and modernity as they are articulated in the places where extraction materializes; 2) the problems and ambiguities surrounding extraction-led development; and 3) new development initiatives and partnerships aimed at overcoming such problems and ambiguities and further propelling development. In all of these domains (which should not be considered as successive phases, but rather as contingent spheres), anthropologists have critically interrogated the politics and the effects of extraction-led development, thereby articulating it with the complex social structures of the lives and lifeworlds in which it assumes form and meaning. In doing so, anthropology has drawn attention to the prominence and importance of extraction's anticipated possibilities of growth and progress, while simultaneously foregrounding that the relation between extraction and development is rather "discordant", to use Gardner's (2012) term. Moreover, if development is seen as an anti-politics machine, rendering apolitical what is deeply political, as Ferguson (1990) has argued, anthropologists' attention to the deeply unequal distribution of benefits (and costs), and its articulation to broader inequalities and skewed power relations has certainly reinserted politics into the debate.

Finally, the questions and debates that anthropology raises will remain critical, not only because extraction will likely continue to feature in development trajectories, but also because new dynamics will potentially redirect the processes, politics, and effects of extraction-led development. For example, how do global sustainability agendas influence the position of extraction as a driver of development? What new resources are required for such green agendas (think of increased demands for lithium which is used in batteries), and where are these resources found, processed, and put to use? What are the geopolitical and local effects of national decisions to continue extracting non-renewable resources to stimulate growth? But also, will alternatives to development – such as those articulated in ideas of de-growth (Escobar 2015) or *Buen Vivir*, an approach which adopts ideas of a "good life" in a broad sense and questions the foundations of modernity (Gudynas 2011) – come to redirect extraction-led development, and if so, how? What will happen when employment opportunities become increasingly limited due to the automation of extractive processes? And what are the effects of global crises such as the COVID-19 pandemic on development ideologies, trajectories, and outcomes (e.g. Hilson et al. 2021; Leach et al. 2021)? Consequently, as new discourses, market dynamics, regulations, policies, technologies, and priorities for stimulating growth and progress continue to emerge, anthropology will also continue to occupy a key role in addressing how extraction-led development is imagined, anticipated, experienced, and negotiated. Studying these dynamics remains key, both to counter hegemonic narratives that uncritically position extraction as a development panacea (or as a devastating curse, for that matter) and, ultimately, to reach a better understanding of how local lives and lifeworlds shape and are shaped by extraction-led development.

Acknowledgements

I would like to thank Lorenzo D'Angelo, Fabiana Li, and Mark Nuttall for their helpful comments on earlier versions of this chapter. The writing of this chapter was financed by the Gold Matters project, which is part of the Belmont Forum and NORFACE Joint Research Programme on Transformations to Sustainability (grant number: 462.17.201) and co-funded by DLR/BMBF, ESRC, FAPESP, ISC, NWO, VR, and the European Commission through Horizon 2020.

Notes

1 Although highly criticized, as the work of Frank, Escobar, and Ferguson, among others, indicates. See beginning of Section 3.
2 See www.sdgs.un.org/goals
3 A distinction is often made between development anthropologists who *work in* development, and anthropologists of development who *study* development (see Escobar 1997). This chapter discusses the work of the latter, although it recognizes that anthropologists also increasingly engage with the extractive industries as practitioners or in policy contexts (see Bainton and Skrzypek this volume).

References

African Union (2009) *Africa mining vision*, Addis Ababa: African Union.

Ahasan A. and Gardner K. (2016) Dispossession by "development": Corporations, elites and NGOs in Bangladesh. *South Asia Multidisciplinary Academic Journal* [Online] 13.

Ahmad N. and Lahiri-Dutt K. (2006) Engendering mining communities: Examining the missing gender concerns in coal mining displacement and rehabilitation in India. *Gender, Technology and Development* 10 (3): 313–339.

Akiwumi F.A. (2012) Global incorporation and local conflict: Sierra Leonean mining regions. *Antipode* 44 (3): 581–600.

Allen T. and Thomas A. (2000) *Poverty and development into the 21st century*, Oxford: Oxford University Press, in association with The Open University.

Appel H. (2012) Walls and white elephants: Oil extraction, responsibility, and infrastructural violence in Equatorial Guinea. *Ethnography* 13 (4): 439–465.

Arce A. and Long N. (2000) Reconfiguring modernity and development from an anthropological perspective. In: Arce A. and Long N. (eds) *Anthropology, development and modernities: Exploring discourses, counter-tendencies and violence*. London & New York: Routledge, 1–31.

Auty R. (1993) *Sustaining development in the mineral economies: The resource curse thesis*, London: Routledge.

Babidge S. (2015) The problem with "transparency": Moral contests and ethical possibilities in mining impact reporting. *Focaal* 73: 70–83.

Bainton N. (2010) *The Lihir destiny: Cultural responses to mining in Melanesia*, Canberra: ANU Press.

Bainton N. (2020) Mining and indigenous peoples. *Oxford research encyclopedia of anthropology* (published online 30 July 2020).

Bainton N. and Jackson R.T. (2020) Adding and sustaining benefits: Large-scale mining and landowner business development in Papua New Guinea. *The Extractive Industries and Society* 7 (2): 366–375.

Banks G., Kuir-Ayius D., Kombako D. and Sagir B. (2013) Conceptualizing mining impacts, livelihoods and corporate community development in Melanesia. *Community Development Journal* 48 (3): 484–500.

Bryceson D.F. and Fisher E. (2014) Artisanal mining's democratizing directions and deviations. In: Bryceson D.F., Fisher E., Jønsson J.B. and Mwaipopo R. (eds) *Mining and social transformation in Africa: Mineralizing and democratizing trends in artisanal production*. London & New York: Routledge. 179–206.

Camacho F.M. (2016) Intergenerational dynamics and local development: Mining and the indigenous community in Chiu Chiu, El Loa Province, northern Chile. *Geoforum* 75: 115–124.

Campbell B. (2009) *Mining in Africa: Regulation and development*, Ottawa: IDRC.

Canavesio R. (2013) Les fronts pionniers des pierres précieuses de Madagascar: Des espaces d'émancipation pour les femmes? *Géocarrefour* 88 (2): 119–129.

Cavalcanti M.T., Da Mata D. and Toscani M.F.G. (2016) *Winning the oil lottery: The impact of natural resource extraction on growth*, Washington, DC: International Monetary Fund.

Chaloping March M. (2017) *Social terrains of mine closure in the Philippines*, London & New York: Routledge.

Cohen R. (2014) Extractive desires: The moral control of female sexuality at Colombia's gold mining frontier. *The Journal of Latin American and Caribbean Anthropology* 19 (2): 260–279.

Coronil F. (1997) *The magical state: Nature, money, and modernity in Venezuela*, Chicago: University of Chicago Press.

Cowen M. and Shenton R.W. (1996) *Doctrines of development*, London: Routledge.

Cross J. (2014) The coming of the corporate gift. *Theory, Culture & Society* 31 (2–3): 121–145.

D'Angelo L. (2019) God's gifts: Destiny, poverty, and temporality in the mines of Sierra Leone. *Africa Spectrum* 54 (1): 44–60.

de Theije, M. (2016) Small-scale gold mining in the Guianas: Mobility and policy across national borders. In: Hoefte R., Bishop M.L. and Clegg P. (eds) *Post-colonial trajectories in the Caribbean*. London & New York: Routledge, 92–106.

Dolan C. and Rajak D. (2016) Introduction: Towards an anthropology of corporate social responsibility. In: Dolan R. and Dinah R. (eds) *The anthropology of corporate social responsibility*. New York: Berghahn Books.

Edelman M. and Haugerud A. (2005) *The anthropology of development and globalization: From classical political economy to contemporary neoliberalism*, Oxford: Wiley-Blackwell.

Escobar A. (1988) Power and visibility: Development and the invention and management of the Third World. *Cultural Anthropology* 3 (4): 428–443.

Escobar A. (1997) Anthropology and development. *International Social Science Journal* 49 (154): 497–515.

Escobar A. (2015) Degrowth, postdevelopment, and transitions: A preliminary conversation. *Sustainability Science* 10 (3): 451–462.

Ferguson J. (1990) *The anti-politics machine: "Development", depoliticization and bureaucratic power in Lesotho*, Cambridge: Cambridge University Press.

Ferguson J. (1999) *Expectations of modernity: Myths and meanings of urban life on the Zambian Copperbelt*, Berkeley, Los Angeles & London: University of California Press.

Ferguson J. (2005) Seeing like an oil company: Space, security and global capital in neoliberal Africa. *American Anthropologists* 107 (3): 377–382.

Ferguson J. (2006) *Global shadows: Africa in the neoliberal world order*, Durham: Duke University Press.

Fisher E., Mwaipopo R., Mutagwaba W., Nyange D. and Yaron G. (2009) "The ladder that sends us to wealth": Artisanal mining and poverty reduction in Tanzania. *Resources Policy* 34: 32–38.

Frank A.G. (1967) *Capitalism and underdevelopment in Latin America*, New York: NYU Press.

Fraser A. and Larmer M. (2010) *Zambia, mining, and neoliberalism: Boom and bust on the globalized Copperbelt*, New York: Palgrave Macmillan.

Gamu J., Le Billon P. and Spiegel S. (2015) Extractive industries and poverty: A review of recent findings and linkage mechanisms. *The Extractive Industries and Society* 2 (1): 162–176.

Gardner K. (2012) *Discordant development: Global capitalism and the struggle for connection in Bangladesh*, London: Pluto Press.

Gardner K., Ahmed Z., Rana M.M. and Bashar F. (2014) Field of dreams: Imagining development and un-development at a gas field in Sylhet. *South Asia Multidisciplinary Academic Journal*. DOI: doi:10.4000/samaj.3741.

Geenen S. (2018) Underground dreams. Uncertainty, risk and anticipation in the gold production network. *Geoforum* 91: 30–38.

Geschiere P. and Nyamnjoh F. (2000) Capitalism and autochthony: The seesaw of mobility and belonging. *Public Culture* 12 (2): 423–452.

Gilberthorpe E., Agol D. and Gegg T. (2016) "Sustainable mining"? Corporate social responsibility, migration and livelihood choices in Zambia. *The Journal of Development Studies* 52 (11): 1517–1532.

Gilberthorpe E and Banks G. (2012) Development on whose terms?: CSR discourse and social realities in Papua New Guinea's extractive industries sector. *Resources Policy* 37 (2): 185–193.

Gilberthorpe E. and Rajak D. (2017) The anthropology of extraction: Critical perspectives on the resource curse. *The Journal of Development Studies* 53 (2): 186–204.

Golub A. and Rhee M. (2013) Traction: the role of executives in localising global mining and petroleum industries in Papua New Guinea. *Paideuma* 59: 215–236.

Gomez E.T. (2003) Political business in Malaysia: Party factionalism, corporate development, and economic crisis. In Gomez E.T. (ed.) *Political business in East Asia*. London: Routledge, 98–130.

GoSL (2013) *The agenda for prosperity: Road to middle income status. Sierra Leone's Third Generation Poverty Reduction Strategy Paper (2013–2018)*. Freetown: Government of Sierra Leone.

Gudynas E. (2010) The new extractivism of the 21st century: Ten urgent theses about extractivism in relation to current South American progressivism. *Americas Program Report* 21: 1–14.

Gudynas E. (2011) *Buen Vivir:* Today's tomorrow. *Development* 54 (4): 441–447.

High M.M. (2017) *Fear and fortune: Spirit worlds and emerging economies in the Mongolian gold rush*, Ithaca & London: Cornell University Press.

Hilson G. (2002) Harvesting mineral riches: 1000 years of gold mining in Ghana. *Resources Policy* 28 (1–2): 13–26.

Hilson G. (2004) Structural adjustment in Ghana: Assessing the impacts of mining-sector reform. *Africa Today* 51 (2): 53–77.

Hilson G. and McQuilken J. (2014) Four decades of support for artisanal and small-scale mining in sub-Saharan Africa: a critical review. *The Extractive Industries and Society* 1: 104–118.

Hilson G., Van Bockstael S., Sauerwein T., Hilson A. and McQuilken J. (2021) Artisanal and small-scale mining, and COVID-19 in sub-Saharan Africa: A preliminary analysis. *World Development* 139, article number 105315. DOI: doi:10.1016/j.worlddev.2020.105315.

Horowitz L.S. (2004) Toward a viable independence? The Koniambo Project and the political economy of mining in New Caledonia. *The Contemporary Pacific* 16 (2): 287–319.

ICMM (2016) *Role of mining in national economies: Mining contribution index* (3rd edition), London: International Council on Mining and Metals.

IFC (2014) *Sustainable and responsible mining in Africa – A getting started guide*, Washington, DC: International Finance Corporation, World Bank Group.

Jacka J.K. (2016) Development conflicts and changing mortuary practices in a New Guinea mining area. *The Journal of the Polynesian Society* 125 (2): 133–147.

Jacka J.K. (2018) The anthropology of mining: The social and environmental impacts of resource extraction in the mineral age. *Annual Review of Anthropology* 47 (1): 61–77.

Jaramillo P. (2020) Mining leftovers: Making futures on the margins of capitalism. *Cultural Anthropology* 35 (1): 48–73.

Jønsson J.B. and Bryceson D. (2014) Going for gold: Miners' mobility and mining motivation. In: Bryceson D.F., Fisher E., Jønsson J.B. and Mwaipopo R. (eds) *Mining and social transformation in Africa: Mineralizing and democratizing trends in artisanal production*. London & New York: Routledge, 39–57.

Kirsch S. (2010) Guest editorial: Sustainability and the BP oil spill. *Dialectical Anthropology* 34 (3): 295–300.

Koczberski G. and Curry G.N. (2004) Divided communities and contested landscapes: Mobility, development and shifting identities in migrant destination sites in Papua New Guinea. *Asia Pacific Viewpoint* 45 (3): 357–371.

Komu T. (2019) Dreams of treasures and dreams of wilderness – engaging with the beyond-the-rational in extractive industries in northern Fennoscandia. *The Polar Journal* 9 (1): 113–132.

Kuchler M. and Bridge G. (2018) Down the black hole: Sustaining national socio-technical imaginaries of coal in Poland. *Energy Research & Social Science* 41: 136–147.

Lahiri-Dutt K. (2016) *The coal nation: Histories, ecologies and politics of coal in India*, Farnham: Ashgate.

Leach M., MacGregor H., Scoones I. and Wilkinson A. (2021) Post-pandemic transformations: How and why COVID-19 requires us to rethink development. *World Development* 138. DOI: doi:10.1016/j.worlddev.2020.105233.

Leonard L. (2016) *Life in the time of oil: A pipeline and poverty in Chad*, Bloomington: Indiana University Press.

Lewis D. (2005) Anthropology and development: The uneasy relationship. In: Carrier J.G. (ed.) *A Handbook of Economic Anthropology*, Second Edition. Cheltenham, UK: Edward Elgar Publishing, 472–488.

Luning S. (2012) Processing promises of gold: A minefield of company–community relations in Burkina Faso. *Africa Today* 58 (3): 22–39.

Macintyre M. (2007) Informed consent and mining projects: A view from Papua New Guinea. *Pacific Affairs* 80 (1): 49–65.

Mosse D. (2005) *Cultivating development: An ethnography of aid policy and practice*, London: Pluto Press.

Mosse D. (2013) The anthropology of international development. *Annual Review of Anthropology* 42: 227–246.

Nuttall M. (2013) Zero-tolerance, uranium and Greenland's mining future. *The Polar Journal* 3 (2): 368–383.

Owen J.R. and Kemp D. (2014) "Free prior and informed consent", social complexity and the mining industry: Establishing a knowledge base. *Resources Policy* 41: 91–100.

Owen J.R. and Kemp D. (2015) Mining-induced displacement and resettlement: A critical appraisal. *Journal of Cleaner Production* 87 (Supplement C): 478–488.

Pelon R. and Walser G. (2009) *Mining together: Large-scale mining meets artisanal mining: A guide for action*, World Bank Other Operational Studies 12458. Washington, DC: World Bank Group.

Perreault T. (2015) Performing participation: Mining, power, and the limits of public consultation in Bolivia. *The Journal of Latin American and Caribbean Anthropology* 20 (3): 433–451.

Pijpers R.J. (2016a) Mining, expectations and turbulent times: Locating accelerated change in rural Sierra Leone. *History and Anthropology* 27 (5): 504–520.

Pijpers R.J. (2016b) The politics of localness: Claiming gains in rural Sierra Leone. In: Eriksen T.H. and Schober E. (eds) *Identity destabilised: Living in an overheated world*. London: Pluto Press, 153–170.

Pijpers R.J. (2017) I want to be a millionaire: Survival, trust and deception in Sierra Leone's economy of dreams. In: Rooij Vd (ed.) *Anthropological knowledge on the move: Contributions from MA graduates at Dutch universities*. Amsterdam: Amsterdam University Press, 135–153.

Pijpers R.J. (2020) Lost glory or poor legacy? The past as a political instrument in a Sierra Leonean mining town. *Zeitschrift fur Ethnologie* 145: 109–128.

Rahnema M. and Bawtree V. (1997 [2006]) *The post-development reader*, London & New Zork: Zed Books.

Rajak D. (2006) The gift of CSR: Power and the pursuit of responsibility in the mining industry. In: Visser W., McIntosh M., Middleton C. (eds) *Corporate citizenship in Africa: Lessons from the past; paths to the future*. London: Routledge, 190.

Rajak D. (2011) *In good company: An anatomy of corporate social responsibility*, Stanford: Stanford University Press.

Revette AC. (2017) This time it's different: Lithium extraction, cultural politics and development in Bolivia. *Third World Quarterly* 38 (1): 149–168.

Reyna S. and Behrends A. (2008) The crazy curse and crude domination: Toward an anthropology of oil. *Focaal* 52: 3–17.

Rogers D. (2014) Energopolitical Russia: Corporation, state, and the rise of social and cultural projects. *Anthropological Quarterly* 87 (2): 431–451.

Rostow W.W. (1960 [1990]) *The stages of economic growth: A non-communist manifesto*, Cambridge: Cambridge University Press.

Rubbers B. (2017) Towards a life of poverty and uncertainty? The livelihood strategies of Gécamines workers after retrenchment in the DRC. *Review of African Political Economy* 44 (152): 189–203.

Sawyer S. (2004) *Crude chronicles: Indigenous politics, multinational oil, and neoliberalism in Ecuador*, Durham & London: Duke University Press.

Sawyer S. and Gomez E.T. (2012) *The politics of resource extraction: Indigenous peoples, multinational corporations and the state*, Chippenham: Palgrave Macmillan.

Schilling-Vacaflor A. and Eichler J. (2017) The shady side of consultation and compensation: "Divide-and-rule" tactics in Bolivia's extraction sector. *Development and Change* 48 (6): 1439–1463.

Schreer V. (2020) "Only gold can become hope": Resource rushes and risky conviviality in Indonesian Borneo. *Ethnos*: 1–23. DOI: doi:10.1080/00141844.2020.1743337

Sharp J. (2006) Corporate social responsibility and development: An anthropological perspective. *Development Southern Africa* 23 (2): 213–222.

Smith Rolston J. (2015) Turning protesters into monitors: Appraising critical collaboration in the mining industry. *Society & Natural Resources* 28 (2): 165–179.

Stammler F. and Wilson E. (2006) Dialogue for development: An exploration of relations between oil and gas companies, communities, and the state. *Sibirica* 5 (2): 1–43.

Szablowski D. (2010) Operationalizing free, prior, and informed consent in the extractive industry sector? Examining the challenges of a negotiated model of justice. *Canadian Journal of Development Studies/Revue canadienne d'études du développement* 30 (1–2): 111–130.

Teaiwa K.M. (2014) *Consuming Ocean Island: Stories of people and phosphate from Banaba*, Bloomington: Indiana University Press.

Thomas A. (2000) Poverty and the end of development. In: Allen T. and Thomas A. (eds) *Poverty and development into the 21st century*. Oxford: Oxford University Press, in association with The Open University, 3–22.

Trigger D.S. (1997) Mining, landscape and the culture of development ideology in Australia. *Ecumene* 4 (2): 161–180.

Verbrugge B. (2015) Undermining the state? Informal mining and trajectories of state formation in Eastern Mindanao, Philippines. *Critical Asian Studies* 47 (2): 177–199.

Voyles T.B. (2015) *Wastelanding: Legacies of uranium mining in Navajo country*, Minneapolis: University of Minnesota Press.

Warburton E. (2017) Resource nationalism in Indonesia: Ownership structures and sectoral variation in mining and palm oil. *Journal of East Asian Studies* 17 (3): 285–312.

WCED (1987) *Our common future*. Oxford & New York: Oxford University Press.

Weber-Fahr M., Strongman J., Kunanayagam R., McMahon G. and Shelton C. (2001) *Mining and poverty reduction*. Washington, DC: The World Bank.

Welker M. (2009) "Corporate security begins in the community": Mining, the corporate social responsibility industry, and environmental advocacy in Indonesia. *Cultural Anthropology* 24 (1): 142–179.

Welker M. (2014) *Enacting the corporation: An American mining firm in post-authoritarian Indonesia*, Berkeley & Los Angeles: University of California Press.

Weszkalnys G. (2008) Hope & oil: Expectations in São Tomé e Príncipe. *Review of African Political Economy* 35 (117): 473–482.

Wiegink N. (2018) Imagining booms and busts: Conflicting temporalities and the extraction–"development" nexus in Mozambique. *The Extractive Industries and Society* 5 (2): 245–252.

Yessenova S. (2012) The Tengiz oil enclave: Labor, business, and the state. *PoLAR: Political and Legal Anthropology Review* 35 (1): 94–114.

4

ENVIRONMENTAL CHANGE

Mark Nuttall

1 Introduction

From the phases of resource exploration and the construction and operation of projects, and through the massive infrastructure and planetary networks this involves, minerals, oil, and gas are extracted, transported, processed, circulated, and consumed within complex transnational social, political, and economic systems (e.g., Jacka 2018; Rogers 2015). On many scales, from the temporal to the spatial, and the granular to the global, extractive industries involve subterranean intrusion, geophysical rupture, ecological damage, and environmental ruination. They entangle places, people, economies, and ecosystems. They excavate and move rock and soil. They reconfigure and pollute environments, with impacts on wildlife, land use, marine ecosystems, groundwater, and subterranean depths. They disperse sand, sediment, and dust, produce waste, leave lasting toxic legacies, and contribute enormously to global carbon emissions. So much so that some of the impacts of resource extraction are now argued to be defining features and stratigraphic markers of the Anthropocene (Bridge 2017; Dethier et al. 2018; Hazen et al. 2017; Melo Zurita, Munro, and Houston 2018; Wagreich and Draganits 2018).

As oil, gas, and minerals become depleted, extractive industries expand their reach into places that are often viewed by states and companies as new terrains of resource extraction. In this process, ideas about discovery and imaginaries of the frontier are persistent and seductive in how resource spaces are marked out (Cons and Eilenberg 2019; Nuttall 2017). Increasingly, this includes areas that are seen as "marginal", "remote", and "peripheral" such as in the Arctic, which are given form as "last frontiers" rich in "undiscovered" resources, even though they are homelands for Indigenous and local communities (Nuttall 2010; Stammler and Ivanova 2016). Extractive resource companies are also exploring the prospects for hydraulic fracturing for what were previously seen as unviable and unconventional oil and

DOI: 10.4324/9781003018018-4

gas resources (De Rijke 2017; Willow and Wylie 2014), or are investigating the possibilities for seabed mining (Childs 2020; Zalik 2018). "As key resources become scarce," write Richardson and Weszkalnys (2014: 5),

> new resources come into existence. Across the globe, states and corporations have redoubled efforts to extract conventional and unconventional resources in an attempt to deliver ongoing prosperity to citizens and shareholders. The contradictions and violence of these endeavors are most apparent in state-sanctioned encroachment of multinational companies on indigenous and other rural lands.

This chapter discusses some key themes in anthropological research on resource extraction and environmental change. While not a comprehensive review, I consider the making of resource spaces, Indigenous peoples and extractivist violence, the transnational nature and planetary reach of the environmental consequences of resource extraction, toxic legacies, environmental change, and subterranean entanglements, and conclude with some remarks on anthropological engagement with understanding resource extraction and the Anthropocene. I also draw on work by scholars in other disciplines. Such multidisciplinary scope is essential given that the magnitude of environmental problems facing the planet demands critical and interdisciplinary examination of how capitalism and global livelihoods depend on resource extraction. This has an added urgency at a time when the activities of resource companies are extending further into and below spaces that are thought of as the remotest edges of the globe, disrupting and transforming environments in the process.

2 The making of resource spaces

Resource spaces – the places in which exploration, extraction, and processing occur – are commonly sites of contestation, dispute, violence, and environmental ruin. Anthropologists have shown how the making of such sites, which are grounded in "historical models of European conquest" (Tsing 2005: 27), involves the dynamics of frontiers and territorialization. These processes threaten, subvert, and dissolve existing social worlds and re-order surface and subterranean spaces (Günel 2019; Jacka 2015; Rasmussen and Lund 2018). As transpires below, in writing about how frontiers are made, re-created, and pushed back, anthropologists have been concerned with understanding how places in which resources are defined, located, and made are imagined and classified as remote and pristine, empty and wild, and how they become subject to techniques and practices of conquest, commodification, and export-oriented resource extraction. Anna Tsing argues that:

> frontiers create wildness so that some – and not others – may reap its rewards. Frontiers are unregulated because they arise in the interstitial places made by collaborations among legitimate and illegitimate partners: armies and bandits;

gangsters and corporations; builders and despoilers. They confuse the boundaries of law and theft, governance and violence, use and destruction.

(Tsing 2005: 27)

The complex histories and contemporary natures of such frontier making and the
accompanying social and environmental impacts are traced with nuance and
insight, for example, in Stephen Nugent's historical anthropology of the Amazon
rubber boom (Nugent 2018), and by Tsing's account of logging and mining in the
rainforests of Kalimantan, which she calls a "zone of awkward engagement". For
Tsing the extractive frontier is an imaginative project in a site of transformations
"made in the shifting terrain between legality and illegality, public and private
ownership, brutal rape and passionate charisma, ethnic collaboration and hostility,
violence and law, restoration and extermination" (Tsing 2005: 33). Similarly, Mary
Louise Pratt used the term "contact zones" to describe those spaces where "cultures, meet, clash and grapple with each other, often in contexts of highly asym-
metrical relations of power such as colonialism, slavery, or their aftermaths as they
are lived out in many parts of the world today" (1992: 4). Resource spaces are such
contact zones of encounters, antagonisms, contestations, struggles, and entanglement (Bainton and Owen 2019; Bridge 2004; Pijpers and Eriksen 2018). As
Gómez-Barris (2017) remarks, the extractive zone is where coordinated forms of
capitalist power, technologies, and worldview mark out and reduce places to spaces
of digging out, production, processing, and transformation in which natural resources
are converted into global commodities. Rasmussen and Lund (2018) discuss how the
making of frontiers and the territorialization of resource control involves processes
that, preceding legitimacy and authority, challenge, overturn, and replace existing
patterns of spatial control, authority, and social and institutional order. And as the
essays in Cons and Eilenberg's edited volume *Frontier Assemblages* (2019) make clear,
the processes of extraction and production redefine and incorporate margins and
remote areas into new territorial formations and global networks. Resource frontiers
are "sites of creative, if often ruinous, production" – "frontier assemblages" are "the
intertwined materialities, actors, cultural logics, spatial dynamics, ecologies, and political economic processes that produce particular places as resource frontiers" (Cons
and Eilenberg 2019: 2). While Luning (2018) offers some cautionary and critical
remarks about the resurgence of the frontier concept as an analytical tool in studies of
resource extraction, Cons and Eilenberg argue that:

what matters in the incorporation (or re-incorporation) of margins are the
various forces and processes that are assembled to reinvent these spaces as zones
of opportunity. And second, we suggest that not only are these forces of spatial
transformation resonant across sites, resources, and interventions, but that a
broader view of territorial intervention gives us tools to understand a moment
in which the relationship of millions of people to land and rule is being radically reconfigured.

(Cons and Eilenberg 2019: 2)

In this making of the frontier and resource spaces, along with an array of abstractive, calculative, and speculative practices, extractive industries displace people and wildlife, disrupt biodiversity, pollute land, rivers, lakes, and seas, and destroy habitat and sacred sites (e.g. Fentiman 1996; Fentiman and Zabbey 2015; Karlsson 2011; Kirsch 2014; Li 2015). In the Arctic, for example, tundra and boreal environments have been disturbed, often in violent ways, by extensive industrial development, such as oil extraction facilities, pipelines and trails from seismic surveys, mining projects (abandoned mine sites have their own toxic legacies), or commercial forestry and clear-cut logging (e.g., Herrmann et al. 2014; Keeling and Sandlos 2015; Sirina 2009; Yakovleva 2011). In the Andes, biodiversity is threatened by environmental degradation caused by resource extraction projects – and it is governments which determine how and where these can take place (Bax, Francesconi, and Delgado 2019); Guzmán-Gallegos (2019: 53) remarks how in northern Peruvian Amazonia oil development has "created landscapes of scattered debris"; Jacka (2015) explores how large-scale gold mining has polluted and degraded the environment in the highlands of Papua New Guinea with consequences that are evident in the disruption of social relations, inequality, and violence; Li (2015) analyses the effects of mining on water, farmland, and sacred places, as well as what it means to live with toxic emissions in a smelter town in Peru; Fentiman (1996: 90) describes how oil has transformed the Niger Delta and has had significant social and economic effects on fishing communities since the first production activities and the construction of oil terminals in the late 1950s; and as Willow (2012, 2017) shows in her ethnography of the threat posed by industrial logging to the traditional territory of the Grassy Narrows First Nation in Ontario in Canada, extractive practices not only entail environmental degradation, they also threaten traditional land-based activities and affect social and cultural life. This is strikingly evident in the Arctic, where mining and oil and gas activities have cumulative effects on traditional resource use practices, such as hunting, fishing, and reindeer herding, and on the economies and well-being of Indigenous and local communities (e.g., Chapin et al. 2005).

In an excellent survey of anthropological studies of mining, Jacka (2018: 62) points out that while extractivism has a 500-year history associated with imperialism and colonialism, and is a mode of accumulation whereby raw materials were removed from the Americas, Asia, and Africa to enrich centres of the world economy, it has now become a primary means of socio-economic development pursued by many countries around the world, including post-colonial states and territories (see Nuttall 2017 for an account of Greenland's ambitions for the development of extractive industries as a stated aim towards independence from Denmark). This comes with continued environmental costs. Bebbington et al. (2018), for example, describe how mineral and hydrocarbon extraction and infrastructure have become significant drivers of forest loss, greenhouse gas emissions, and threats to the rights of communities in forested areas of Amazonia, but also in Indonesia and Mesoamerica. Governments there, as in many other parts of the world, view resource extraction as a pathway toward development, alleviating poverty and bringing prosperity, and have made framework policy commitments to increase energy production and the further exploitation of natural resources. Yet Bebbington and colleagues argue that these threats to the environment and to the rights of Indigenous and rural peoples will continue to be substantial, particularly because resource extraction and its associated infrastructure also enable population movements and agricultural expansion further into the forest. Conservation policies that protect forests and the lands of Indigenous and rural people are under threat, small-scale mining has intensified in specific locations and also has become a driver of deforestation and degradation. Forest dwellers' perceptions of insecurity have increased, as have incidences of political violence and the documented homicides of environmental activists (Bebbington et al. 2018).

3 Indigenous peoples, extractivist violence, and environmental change

Ancestral and traditional lands sustain the lifeways of Indigenous peoples, whether they hunt, herd, fish, grow subsistence crops, or practice a combination of land and water-based activities that integrate formal and informal economies. Human–environment relationships emphasizing reciprocity are essential for cultural and economic survival, social identity, and spiritual life and are reflected in rich mythologies, worldviews, and in moral and ethical codes. Colonization processes and extractive capitalism have proceeded without regard for an understanding of how traditional lands are homelands for Indigenous communities; instead they dispossess them from the places that are turned into resource zones. As environmental sociologist Valerie Kuletz (1998) shows with reference to uranium mining and nuclear test sites in parts of Nevada, New Mexico, and California, for example, extractive spaces and toxic waste dumps are situated in or close to Indigenous lands that are viewed as being in empty wastelands. Indigenous perspectives on human–environment relations are ignored, silenced, and subject to mechanisms of exclusion. Because land is often seen as empty by colonizers and settlers, this view "violates fundamental territorial and cultural rights and aspirations of indigenous peoples" (Dahl 1993: 103). It is for this reason that land and resource rights are often at the very heart of Indigenous discourses on self-determination and self-government.

In September 2020, the CEO and two other senior executives of global mining company Rio Tinto resigned following pressure from investors over the company's decision to blow up 46,000-year-old rock shelters sacred to Aboriginal traditional holders at Juukan Gorge in Western Australia's Pilbara region. The company had destroyed the shelters four months earlier so that it could mine iron ore, although it had known about their cultural and spiritual significance for several years. Initially, Rio Tinto's board of directors had responded by withholding the bonuses of the executives, but investors felt they should be held to account in a way that did not just penalize them financially. This controversy is yet one more example in a long history of violence enacted by extractive industries on environments and

Indigenous peoples. Such violence may often be associated with extractivist expansion (cf. Watts 2001), but is also manifest as epistemic violence in the way that scientific knowledge is used to design and legitimize resource development projects (e.g., Arach 2018; Gómez-Barris 2017; Kuletz 1998; Preston 2017). It is not only Indigenous peoples who experience this violence as well as environmental ruin, although they are often affected disproportionately.

The destruction of the Juukan Gorge shelters brings to mind Werner Herzog's 1984 film *Where the Green Ants Dream*, which opens with a mining company setting off explosions in the Australian outback as part of seismic exploratory activity in search of likely sites to dig for uranium. At the beginning of the film, several Aboriginal elders sit in front of the bulldozers and bring a halt to the mining company's activities. They explain that the area is a place where green ants come to dream and that they should not be woken up. A company executive flies in to talk with the elders and attempts to negotiate and get them to leave the area. The stories and knowledge of the elders are dismissed, however, in a way that acts to privilege the expertise of the company and its mining engineers and geologists over Indigenous relational ontology. The film corresponds with anthropological work that explores how people perceive and experience disturbance from resource exploration and exploitation. In northern Greenland, for example, Inuit hunters talk about *qullugiarsuaq*, a great sea worm believed to live in coastal waters, being unsettled and restless following marine seismic surveys (Nuttall 2017), while Gardner (2007) reports from northern Canada on Dene stories about underground beings that have been disturbed by gas-drilling crews. As D'Angelo (2014) argues with reference to the ways in which people in Sierra Leone sometimes interpret unexpected and mysterious events occurring at a large-scale mining company in terms of the occult, the emergent narratives about such events connected to extractive intrusions into the realm of the non-human are, in fact, often highly politicized practices.

Reports of conflicts between Indigenous communities and nation-states and between Indigenous communities and national and multinational corporations bring into focus concerns over the social, cultural, economic, and environmental impacts of extractive industries on Indigenous peoples, but they also raise the question of who has rights over subsurface resources and to their development. This is particularly so in areas where Indigenous peoples are fighting for land claims and rights over resources and in places where ownership of subsurface rights is contested. In western Canada, for example, the construction and operation phases of oil and gas pipelines affect rural and First Nations communities in a number of ways, disturbing farmland or impeding access to traditional hunting and fishing areas, with consequences for food security and community well-being (Jonasson et al. 2019; Nuttall 2010, 2014). In Alberta, enormous deposits of oilsands continue to be mined in the boreal northlands within the traditional territories of First Nations and Métis communities. Their concerns relate to environmental disturbance, air and water pollution, and habitat loss, as well as the impact of oilsands activities on human health and community well-being. For Jen

Preston, concepts of race and the realities of racism structure the ways the Canadian nation-state and the extractive industries profit from the oilsands. Introducing the idea of racial extractivism, she argues that race, racism, and racialization need to be analysed in order to understand how colonial violence and its extractive manifestations on people and environment are enabled (Preston 2017).

4 Extractivism and the infringement of rights

Drawing attention to how ever-increasing segments of the world's population now experience and have to contend with environmental challenges that they did not authorize, or benefit from, Willow (2014) calls for anthropological investigation of environmental change and degradation in unexpected places, with deep analysis of the implications resource extraction has not just for the environment, but for social and economic inequity and for understanding claims for global sustainability. With respect to Indigenous peoples, Article 26 of the United Nations Declaration on the Rights of Indigenous Peoples (UNDRIP), which was adopted by the UN General Assembly on 13 September 2007, states that "Indigenous peoples have the right to the lands, territories and resources which they have traditionally owned, occupied or otherwise used or acquired." Article 32 further asserts that:

> States shall consult and cooperate in good faith with the indigenous peoples concerned through their own representative institutions in order to obtain their free and informed consent prior to the approval of any project affecting their lands or territories and other resources, particularly in connection with the development, utilization or exploitation of mineral, water or other resources.

Central to this is the principle of free, prior, and informed consent (see also Li and Velásquez, and Pijpers this volume). Yet, around the world, examples of how this does not happen, and how Indigenous rights are infringed, are numerous. For example, Minter et al. (2012) showed how free, prior, and informed consent has been recognized in the Philippines under the Indigenous Peoples' Rights Act of 1997, but its implementation has often failed in terms of both process and outcome. They argue that consent is manipulated and that agreements made between companies and Indigenous groups have been, in many instances, culturally inappropriate and weakly operationalized. In Russia, the pollution generated by the oil and gas industries affects reindeer pasture and marine and freshwater environments. In Siberia and eastern Russia, oil and gas extraction and mining ventures have removed extensive tracts of lands and territories from Indigenous use, and the transport of hydrocarbons across Indigenous lands also impacts traditional activities. Pipelines and facilities create obstacles to the movement and migration of reindeer herds across tundra landscapes. In the Yamal region, for example, the destruction of vegetation due to facilities, road, and pipeline construction has meant that reindeer herds have been concentrated into smaller areas of pasture. This has led to over-grazing, with potential long-term adverse effects on ecosystem productivity and local economies (Forbes 1999). Fondahl and Sirina (2006) discuss how, for much of the late twentieth century, hydrocarbon development in Russia took place on native lands, largely in the western Siberian oil fields on Khanty and Mansi traditional territory and the northwestern gas fields on Nenets homelands. In Russia's far east, controversies over the exploitation of Sakhalin Island's oil reserves have challenged the territorial rights of the Nivkhi, Evenki, and Uilta. While in some cases Russia's Indigenous peoples have received benefits from oil and gas development, they do not have rights to subsurface resources, and generally the costs and the negative impacts of development on society, culture, environment, and wildlife outweigh whatever positive experiences can be identified and chronicled.

Following its acquisition of Burlington Resources in 2006, ConocoPhillips became one of the most significant international energy companies involved in developing oil and gas resources in the Amazon. In Peru alone, it acquired drilling rights in areas covering 10.5 million acres of tropical rainforest (Anderson et al. 2009). In Ecuador, Shuar, Kichwa, and Achuar peoples called for the protection of the rainforest in areas marked off for oil extraction. Elsewhere in the Amazon, Indigenous peoples also attempted to resist the incursions of ConocoPhillips and other multinational companies. In May 2008, the Peruvian government sent troops to back up police who were trying to quell protests by Indigenous peoples over land, oil, and mineral rights in the Maranon River basin. Indigenous peoples' organizations in the country accused the Peruvian government of selling those rights to foreign companies and pointed to a failure to consult with Indigenous communities about plans for resource extraction. The situation illustrated how, from the point of view of many governments, oil and mineral rights are vested in the state and that Indigenous communities are regarded as being an obstacle to extractive industry. Arsel, Hogenboom, and Pellegrini (2016) describe how an "extractive imperative" has arisen in South American countries where natural resource extraction has come to be seen simultaneously as a source of income and employment generation and financing for increased social policy expenditure. According to this imperative, extraction needs to continue and expand regardless of prevailing circumstances, with the state playing a leading role and capturing a large share of the ensuing revenues. However, a wider global movement is emerging in which people's contestations seek to open up new discussions on Indigenous ontologies in places that are affected by extractivism (e.g., Muller, Hemming, and Rigney 2019; Rivera Andia and Vindal Ødegaard 2019). In the face of pressure from extractive industries as well as rapid climate change in Greenland and Canada's Arctic, for instance, Inuit are advancing arguments for collaborative governance initiatives that include indigenous knowledge as a basis for managing marine protected areas (e.g., Inuit Circumpolar Council 2017; Nuttall, 2016, 2021).

5 The transnational and planetary scale of extractive infrastructures

The environmental footprint of mining and oil and gas production is not just confined to geographically bounded locations that have been marked out as resource regions within countries. Extractive infrastructures involve flows and mobilities, as well as economic and financial networks and patterns of work and employment, that stretch around the globe (e.g., Arboleda 2020; Hein 2018). Building on Mazzan Labban's idea of the planetary mine (Labban 2014), Martín Arboleda draws on his own research on mining in Chile's Atacama Desert to show how geographies of extraction are "entangled in a global apparatus of production and exchange that supersedes the premises and internal dynamics of a proverbial world system of cores and peripheries defined exclusively by national borders" (Arboleda 2020: 5). The point here is that a mine is not a "discrete sociotechnical object but a dense network of territorial infrastructures and spatial technologies widely distributed across space" (ibid.). The same can be said of an oil or gas field.

Extractive industry requires a vast infrastructure and associated technologies, as well as the bureaucracies of finance, sales, and export, beyond a mine or oilfield, and extending beyond encampments, boomtowns, and the communities that grow up around extraction. Once coal or iron ore, for instance, have been extracted from the ground, they have to be transported via railways and roads, or by way of harbours and ships to terminals, production and processing facilities, and on to storage facilities and markets. Oil and gas production includes wells, rigs, gathering facilities, pipelines, pump stations, processing plants, and refineries. All of this infrastructure and technology and the transnational connections and labour markets they rely on, the transnational practices and networks they establish, and the patterns of global consumption they depend on, have an environmental impact beyond that of the extraction of the resource.

Pipelines, for example, are not merely single lines of buried steel pipe that run from place of origin delivering oil or gas to a terminus. They are components of vast, complex technological systems. Writing about the United States–Canada natural gas industry, Arthur Mason (2012: 80) describes it as a "large technical system consisting of a vast continental-sized machine made up of 2 million miles of steel pipe". Because this interconnected system of pipelines is buried, Mason says that only a few people are aware of or witness the scale of the techno-ontological dimension of this network first hand, but "the steady stream of fuel to consumers and occasional explosion that destroys lives and entire neighbourhoods" is testimony to the very real material and political presence pipelines have (ibid.). Hein (2018) explores one dimension of this with reference to the relationship between oil refineries and port cities, their entanglement with global events and national strategies, and their spatial and environmental impacts. Yet as Stuart Kirsch argues, given how large-scale resource extraction projects are more often than not dominated by international corporations and distant capital and are primarily responsive to global markets, mining companies are discouraged from "making sufficient investments in environmental controls" (Kirsch 2014: 15).

Building on James Ferguson's (2005) discussion of extractive enclaving, a process that entails little or no economic, social, or cultural benefit for the places in which extraction takes place or for the wider society, anthropologists have argued that extractive practices impact the daily lives and surroundings of people who live not so much on extractive frontiers or zones, but in extractive enclaves (e.g., see Appel 2012 and Rubbers 2019 for an examination of this in Equatorial Guinea and the Congolese copperbelt, respectively). The extractive enclave arises from extractive practices that aim to maximize profit, whatever the social and environmental cost. Discussing various contestations over the ownership of oil in the Niger Delta and how local communities do not benefit, Omolade Adunbi (2020) points out that the extractive enclave is shaped by a global order that connects local and national actors with transnational practices and networks that enable the emergence of alliances and arrangements that privilege the state, oil corporations, and business initiatives over local communities and environments.

Espig and de Rijke (2018) argue that global energy networks prompt anthropologists to rethink the nature of their "field sites" and how they can pay attention to, and understand, extractive industries and their wider reach and social, economic, and environmental impacts in relation to global interconnectedness. As Hein (2018) puts it, constellations of oil actors, which include corporations and nations, shape the physical spaces that appear to be disconnected and geographically distant, but which emerge as a global palimpsestic petroleumscape (see also, Ferguson 2005). In a similar vein, Schritt and Behrends (2018: 212) suggest that oil extraction in West Africa is located in "oil zones" that are trans-territorial spaces. They point to the

> (dis)connections between (distant) places through infrastructures, technologies, borders, fences and security practices, as well as discursive movements of ideas, theories, models and narratives. The concept of "oil zones" thus enables us to look at specific places in Niger and Chad, like the oil extraction sites that are disconnected from their immediate surroundings by fences and military guards, but at the same time connected with distant places in Africa and beyond (United States, Europe, China) through pipelines, specific means of transport and the conversion of crude oil into new products. It also helps us to analyse the particular and quite different forms of (dis)connection between places on a global scale in relation to the ideas, scientific theories and economic models that bring forth these very (dis)connections.

As Mezzadra and Neilson (2017) argue, however, the intensification and expansion of extractive industries in contemporary capitalism demands an attentiveness not just to the literal productive, earth moving, and digging that extraction involves but to the new forms of extraction that are happening on different kinds of frontier such as data mining and biocapitalism. Getting to grips with this involves understanding what they call operations of capital as a way of tracing a complexity of dense connections between the far-reaching, expansive logic of extraction and capitalist activity in the spheres of logistics and finance.

6 Resource extraction, toxic legacies, and subterranean entanglements

The environmental, social, and health impacts of mining and oil and gas extraction, as well as transport, refining, and consumption, are widely distributed and have a diversity of social and environmental impacts which often highlight existing socio-economic, ethnic, and racial inequalities (e.g., see O'Rourke and Connolly 2003). As major accidents such as the Deepwater Horizon oil blowout in the Gulf of Mexico in April 2010 show, the environmental and human impacts can be substantial across a range of scales (Joye 2015). Spills from marine transportation or offshore oil platforms have the potential for widespread and long-lasting ecological damage, particularly in areas that appear to be more vulnerable to the effects of industrial activities, such as ice-covered Arctic waters (Shapavalova-Krout 2019). However, Bond (2013) cautions against viewing oil spills and the environment as given objects that collide during a disaster – rather, both crude oil and the environment are made politically operable in relation to one another. Discussing the Deepwater Horizon containment and clean-up operation, he argues that the sprawling mess caused by the blowout was made into a manageable problem through the transformation of the Gulf of Mexico into a scientific laboratory within which the true size and scope of the hydrocarbon extraction industry could be understood, mastered, and controlled. For Bond, this objectified the oil spill, but it also created the idea of an easily observed and knowable environment in which the disaster could be contained through techniques of sequestering, inspecting, and monitoring the oil spill. The implication of this, he suggests, was to create ways of disconnecting the disaster from the public.

In a number of countries, the reclamation of derelict mining sites has become a key policy objective for environmental management. This is not always a case of just cleaning up the disorder, messiness, and contamination left behind. Habitat can be recreated on ruined, derelict, or degraded sites, as Seitz, van Engelsdorp, and Leonhardt (2019) show for bee conservation in reclaimed sand mines in Maryland in the United States, or can emerge from the ruination. Disused mining landscapes can become experiments in rewilding, or they become forested and hold out hope for conservationists that animals can wander in by themselves, in the way that Tsing (2017) recounts of red deer, biodiversity, and the emergence of a novel landscape in a former brown coal mine site in Denmark. In former Welsh lead mining areas, shafts, tunnels, ledges, and crevasses in abandoned mines, quarries, and gravel pits provide undisturbed breeding and nesting places for some bird species and amphibians, while metalliferous soils and spoil heaps have been colonized by metal-resistant plants. These post-mining landscapes have become redesignated. No longer viewed as ruined by extractive industry, they now have significant conservation status and comprise sites of special scientific interest that are protected for the rare plants and habitats that now characterize them. Yet the toxic legacy of contamination persists, hidden in the soil and the water (Nuttall 2020). Abandoned and closed mine sites also generate new social uses and meanings. For example, Pijpers (2018) shows how alterations to the landscape made by iron ore mining in Sierra Leone are taken up by local actors and feature in the negotiation of ownership, access, and control, while Robertson (2006) focuses on the importance of past mining towns in parts of the United States for nurturing a sense of place and identity. Rather than being emblematic of de-industrialization, dereliction, and decay, some historic mining towns and their ruins, spoil heaps, and rusting machinery "stand as curious relics of a romanticized frontier age" in landscapes perceived as important heritage sites (Robertson 2006: 6).

Resource extraction projects are often relatively temporary, even if they continue for a few decades, but once a mine or an oilfield closes after exhausting the resource, it nonetheless retains a presence (White 2013). This is not just apparent in the physical form of abandoned harbours, roads, airstrips, sheds, machinery, and tailings, or deep below ground in shafts, mines, tunnels, and wells (and notwithstanding reclamation and remediation, evidence of this material presence often remains), but in community memories of extractive activities and their impacts, or in the toxic material that lies deep in contaminated ground or polluted lakes and rivers (Kuletz 1998; Ruggiero and South 2013; Voyles 2015). Lurking within, this toxicity can seep slowly to the surface and enter human bodies, fish, and animals (e.g., Sandlos and Keeling 2016), or can be inhaled as Mazzeo (2018) illustrates in her study of the effects of the bio-persistence of asbestos fibres in the air on the health of people living near an abandoned mine in north-eastern Brazil. Benadusi (2018) describes how the effects of petrochemical industrialization in a Mediterranean area in southeastern Sicily have come to define it as the triangle of death. She argues that oil represents both a blessing and a curse for local people, many of whom feel that everyday forms of catastrophe are present in air pollution or in long-term health problems, generating effects that can appear far more insidious than a major disaster such as an oil spill or refinery fire. Discussing two major oil spills in the Niger Delta in 2008, Fentiman and Zabbey (2015) show that while the environmental degradation was visible in creeks and swamps, the social impacts and cultural erosion experienced by the Bodo community in Ogoniland were less so. In their discussion of the legacy of the North Rankin Nickel Mine at Rankin Inlet in Nunavut in Canada, Cater and Keeling (2013) discuss how tailings and other toxic elements are felt by residents to have a malign presence in the landscape. At the same time, while they are concerned about the environmental legacies of the mine, people engage with the mining past in ways that stimulate forms of remembering of the positive aspects of extractive activity and its importance for social and economic life. As Cater and Keeling (2013: 64) put it, "the material remnants of mineral extraction link past events and present encounters through ongoing and everyday encounters, as people come into contact with material objects". Equipment, machinery, old buildings, and mine shafts are repositories of community memories, significant for social and cultural identity and for thinking about a future in a place that is also likely to be characterized by resource extraction as industry continues to cast its gaze on the surrounding area.

7 Conclusions

In Book XXXIII of *Natural History*, written and compiled in the eighth decade of the first century AD, Pliny the Elder reflected on the practice and effects of extractive industry in the Roman Empire. He wrote how, in some places, "the earth is dug into for riches", while in other places it is dug into "for rash valour", with iron being prized more than gold because of its use in the manufacture of weapons. "We descend into the bowels of the earth; and seek for wealth even in the abodes of departed spirits," he stated, "trace out all the fibres of the earth, and live above the hollows we have made in her." Pliny's words seem apt today when we reflect on a global history of resource extraction and how humans have sculpted, transfigured, disrupted, and modified the nature and properties of the Earth's surface and subsurface. In this chapter, I have considered how anthropologists and other scholars have contributed to understanding extraction and environmental change in relation to the making of resource spaces, extractivist violence, the global reach of resource extraction, toxic legacies, and subterranean entanglements. In bringing it to a conclusion, it is important to note that the lingering presence and the environmental legacies of extractive industry – even from some of the world's earliest mining ventures (e.g., see Wagreich and Draganits 2018) – are increasingly becoming known through research seeking to discover and define stratigraphic, sedimentary, and sequential markers as evidence for the Anthropocene. Anthropologists are reflecting on how they can contribute to these investigations (e.g., Mathews 2020).

Research agendas are taking shape that highlight the potential of anthropological engagement with the ways environmental change affects the subterranean depths and spaces into which extractive industries probe and dig and from which they unearth minerals and hydrocarbons. Some of this thinking is informed by approaches to territory, political geology, elemental geographies, and deep time, which have drawn attention to how thinking vertically and volumetrically about space and place, as well as what D'Angelo and Pijpers (2018) call mining temporalities, enlivens our understanding of subterranean environments, the human impacts on them and, in turn, the implications for social life (e.g., Bobbette and Donovan 2019; Elden 2013; Hawkins 2020; Irvine 2020; see also Luning this volume). Vertical and volumetric perspectives have much to offer to anthropology, especially with respect to how the subterranean has been imagined, made legible, and controlled, and how resource potential is anticipated, assessed, and calculated.

A challenge for anthropology – and for its broader concern with the Anthropocene – will be to contribute to understanding how the human impacts of extraction are revealed to have left their deep and lasting traces on Earth processes and geologies. To do this, anthropologists must play a major role in multidisciplinary and interdisciplinary efforts that enrich our knowledge of the ways humans are entangled with Earth's geological and temporal depths.

Bibliography

Adunbi O. (2020) Extractive practices, oil corporations and contested spaces in Nigeria. *The Extractive Industries and Society* 7 (3): 804–811.

Anderson M., Finer M., Herriges D., Miller A. and Soltani A. (2009) *ConocoPhillips in the Peruvian Amazon.* A report by Amazon Watch and Save America's Forests.

Appel H. (2012) Walls and white elephants: Oil extraction, responsibility, and infrastructural violence in Equatorial Guinea. *Ethnography* 13 (4): 439–465.

Arach O. (2018) Like an army in enemy territory: Epistemic violence in megaextractivist expansion. In: Oswald Spring Ú. and Serrano Oswald S.E. (eds) *Risks, security, violence and peace in Latin America.* Cham: Springer, 100–111.

Arboleda M. (2020) *Planetary mine: Territories of extraction under late capitalism*, London and New York: Verso.

Arsel M., Hogenboom B. and Pellegrini L. (2016) The extractive imperative in Latin America. *The Extractive Industries and Society* 3 (4): 880–887.

Bainton N. and Owen J.R. (2019) Zones of entanglement: Researching mining arenas in Melanesia and beyond. *The Extractive Industries and Society* 6 (3): 767–774.

Bax V., Francesconi W. and Delgado A. (2019) Land use conflicts between biodiversity conservation and extractive industries in the Peruvian Andes. *Journal of Environmental Management* 232: 1028–1036.

Bebbington A.J., Humphreys-Bebbington D., Sauls L.A., Rogan J., Agrawal S., Gamboa C. et al. (2018) Resource extraction and infrastructure threaten forest cover and community rights. *PNAS* 115 (52): 13164–13173.

Benadusi M. (2018) Oil in Sicily: Petrocapitalist imaginaries in the shadow of old smoke-stacks. *Economic Anthropology* 5 (1): 45–58.

Bobbette A. and Donovan A. (eds) (2019) *Political geology: Active stratigraphies and the making of life*, Cham: Palgrave MacMillan.

Bond D. (2013) Governing disaster: The political life of the environment during the BP oil spill. *Cultural Anthropology* 28 (4): 694–715.

Bridge G. (2004) Contested terrain: Mining and the environment. *Annual Review of Environment and Resources* 29: 205–259.

Bridge G. (2017) Resource extraction. In Richardson D. (ed.) *International encyclopedia of geography: People, the earth, environment and technology.* Wiley online.

Cater T. and Keeling A. (2013) "That's where our future came from": Mining, landscape, and memory in Rankin Inlet, Nunavut. *Études/Inuit/Studies* 37 (2): 59–82.

Chapin S., Berman M., Callaghan T.V., Convey P., Crépin A-S., Danell K. et al. (2005) Polar systems. In: Millennium Ecosystem Assessment *Ecosystems and human well-being: Conditions and trends.* Washington DC: Island Press, 717–743.

Childs J. (2020) Extraction in four dimensions: Time, space and the emerging geo(-)politics of deep-sea mining. *Geopolitics* 25 (1): 189–213.

Cons J. and Eilenberg M. (2019) Introduction: On the new politics of margins in Asia: Mapping frontier assemblages. In: Cons J. and Eilenberg M. (eds) *Frontier assemblages: The emergent politics of resource frontiers in Asia.* Oxford: Wiley, 1–18.

D'Angelo L. (201) Changing environments, occult protests, and social memories in Sierra Leone. *Social Evolution and History* 13 (2): 22–56.

D'Angelo L. and Pijpers R.J. (2018) Mining temporalities: An overview. *The Extractive Industries and Society* 5 (2): 215–222.

Dahl J. (1993) *Indigenous peoples of the Arctic*. In: Arctic Challenges. Report from the Nordic Council Parliamentary Conference in Reykjavik, Iceland, 16–17 August.

De Rijke K. (2017) Produced water, money water, living water: Anthropological perspectives on water and fracking. *WIREs Water* 5 (2): e1272.

Dethier D.P., Ouimet W.B., Murphy S.F., Kotikian M., Wicherski W. and Samuels R.M. (2018) Anthropocene landscape change and the legacy of nineteenth- and twentieth-century mining in the Fourmile Catchment, Colorado Front Range. *Annals of the American Association of Geographers* 108 (4): 917–937.

Elden S. (2013) Secure the volume: Vertical geopolitics and the depth of power. *Political Geography* 34: 35–51.

Espig M. and de Rijke K. (2018) Energy, anthropology and ethnography: On the challenges of studying unconventional gas developments in Australia. *Energy Research and Social Science* 45: 214–223.

Fentiman A. (1996) The anthropology of oil: The impact of the oil Industry on a fishing community in the Niger Delta. *Social Justice* 23 (4): 87–99.

Fentiman A. and Zabbey N. (2015) Environmental degradation and cultural erosion in Ogoniland: A case study of the oil spills in Bodo. *The Extractive Industries and Society* 2 (4): 615–624.

Ferguson J. (2005) Seeing like an oil company: Space, security, and global capital in neoliberal Africa. *American Anthropologist* 107 (3): 377–382.

Fondahl G. and Sirina A. (2006) Rights and risks: Evenki concerns regarding the proposed Eastern Siberia–Pacific Ocean Pipeline. *Sibirica* 5 (2): 115–138.

Forbes B.C. (1999) Land use and climate change in the Yamal-Nenets region of northwest Siberia: Some ecological and socio-economic considerations. *Polar Research* 18 (2): 367–373.

Gardner P.M. (2007) On puzzling wavelengths. In: Goulet J-G. and Miller B-G. (eds) *Extraordinary anthropology: Transformations in the field*. Lincoln and London: University of Nebraska Press.

Gómez-Barris M. (2017) *The extractive zone: Social ecologies and decolonial perspectives*, Durham, NC: Duke University Press.

Günel G. (2019) Subsurface workings: How the underground becomes a frontier. In: Cons J. and Eilenberg M. (eds) *Frontier assemblages: The emergent politics of resource frontiers in Asia*. Oxford: Wiley, 41–57.

Guzmán-Gallegos M.A. (2019) Controlling abandoned oil installations: Ruination and ownership in northern Peruvian Amazonia. In: Vindal Ødegaard C. and Rivera Andía J.J. (eds) *Indigenous life projects and extractivism: Approaches to social inequality and difference*. Cham: Springer, 53–73.

Hawkins H. (2020) Underground imaginations: Environmental crisis and subterranean cultural geographies. *Cultural Geographies* 27 (1), 3–22.

Hazen R.M., Grew E.S., Origlieri M.J. and Downs R.T. (2017) On the mineralogy of the "Anthropocene Epoch". *American Mineralogist* 102: 595–611.

Hein C. (2018) "Old refineries rarely die": Port city refineries as key nodes in the global petroleumscape. *Canadian Journal of History* 53 (3): 450–479.

Herrmann T.M., Sandström P., Granqvist K., D'Astous N., Vannar J., Asselin H. et al. (2014) Effects of mining on reindeer/caribou populations and indigenous livelihoods: Community-based monitoring by Sami reindeer herders in Sweden and First Nations in Canada. *The Polar Journal* 4 (1): 28–51.

Inuit Circumpolar Council (2017) *People of the Ice Bridge: The future of the Pikialasorsuaq*, Report of the Pikialasorsuaq Commission, Ottawa: Inuit Circumpolar Council Canada.

Irvine R.D.G. (2020) *An anthropology of deep time: Geological temporality and social life*, Cambridge: Cambridge University Press.

Jacka J.K. (2015) *Alchemy in the rain forest: Politics, ecology, and resilience in a New Guinea mining area*, Durham, NC: Duke University Press.

Jacka J.K. (2018) The anthropology of mining: The social and environmental impacts of resource extraction in the mineral age. *Annual Review of Anthropology* 47: 61–77.

Jonasson M.E., Spiegel S.J., Thomas S., Yassi A., Wittman H., Takaro T. et al. (2019) Oil pipelines and food sovereignty: Threat to health equity for Indigenous communities. *Journal of Public Health Policy* 40: 504–517.

Joye S.B. (2015) Deepwater Horizon, 5 years on. *Science* 349 (6248): 592–593.

Karlsson B.G. (2011) *Unruly hills: A political ecology of India's Northeast*, New York and Oxford: Berghahn Books.

Keeling A. and Sandlos J. (2015) *Mining and communities in Canada: History, politics and memory*, Calgary: University of Calgary Press.

Kirsch S. (2014) *Mining capitalism: The relationship between corporations and their critics*, Oakland, CA: University of California Press.

Kuletz V.L. (1998) *The tainted desert: Environmental and social ruin in the American West*, New York and London: Routledge.

Labban M. (2014) Deterritorializing extraction: Bioaccumulation and the planetary mine. *Annals of the Association of American Geographers* 104 (3): 560–576.

Li F. (2015) *Unearthing conflict: Corporate mining, activism, and expertise in Peru*, Durham, NC: Duke University Press.

Luning S. (2018) Mining temporalities: Future perspectives. *The Extractive Industries and Society* 5 (2): 281–286.

Mason A. (2012) Industry as Alaska's ethnographic present. *The Polar Journal* 2 (1): 77–92.

Mathews A.S. (2020) Anthropology and the Anthropocene: Criticisms, experiments, and collaborations. *Annual Review of Anthropology* 49: 67–82.

Mazzeo A. (2018) The temporalities of asbestos mining and community activism. *The Extractive Industries and Society* 5 (2): 223–239.

Melo Zurita M., Munro P.G. and Houston D. (2018) Unearthing the subterranean Anthropocene. *Area* 50 (3): 298–305.

Mezzadra S. and Neilson B. (2017) On the multiple frontiers of extraction: Excavating contemporary capitalism. *Cultural Studies* 31 (2–3): 185–204.

Minter T., de Brabander V., van der Ploeg J., Persoon G.A. and Sutherland T. (2012) Whose consent? Hunter-gatherers and extractive industries in the Northeastern Philippines. *Society and Natural Resources* 25 (12): 1241–1257.

Muller S., Hemming S. and Rigney D. (2019) Indigenous sovereignties: Relational ontologies and environmental management. *Geographical Research* 57 (4): 399–410.

Nugent S.L. (2018) *The rise and fall of the Amazon rubber industry: An historical anthropology*, London and New York: Routledge.

Nuttall M. (2010) *Pipeline dreams: People, environment and the Arctic energy frontier*. Copenhagen: IWGIA.

Nuttall M. (2014) Pipeline politics in northwest Canada. In: Powell R.C. and Dodds K. (eds) *Polar geopolitics? Knowledges, resource and legal regimes*. Cheltenham: Edward Elgar, 277–294.

Nuttall M. (2016) Narwhal hunters, seismic surveys and the Middle Ice: Monitoring environmental change in Greenland's Melville Bay. In: Crate S.A. and Nuttall M. (eds) *Anthropology and climate change: From actions to transformations*. London and New York: Routledge, 354–372.

Nuttall M. (2017) *Climate, society and subsurface politics in Greenland: Under the Great Ice*, London and New York: Routledge.

Nuttall M. (2020) Lead mining, conservation and heritage: Shaping a mountain in northeast Wales. *Humanities* 9 (3): 70.

Nuttall M. (2021) Arctic ecology, Indigenous peoples and environmental governance. In: Thomas D. (ed.) *Arctic ecology*. Oxford: Wiley Blackwell, 409–422.

O'Rourke D. and Connolly S. (2003) Just oil? The distribution of environmental and social impacts of oil production and consumption. *Annual Review of Environment and Resources* 28: 587–617.

Pijpers R.J. (2018) Territories of contestation: Negotiating mining concessions in Sierra Leone. In: Pijpers R.J. and Eriksen T.H. (eds) *Mining encounters: Extractive industries in an overheated world*. London: Pluto Press, 78–96.

Pijpers R.J. and Eriksen T.H. (2018) *Mining encounters: Extractive industries in an overheated world*. London: Pluto Press.

Pratt M.L. (1992) *Imperial eyes: Travel writing and transculturation*, London: Routledge.

Preston J. (2017) Racial extractivism and white settler colonialism: An examination of the Canadian Tar Sands mega-projects. *Journal of Cultural Studies* 31 (2–3):353–375.

Rasmussen M.B. and Lund C. (2018) Reconfiguring frontier spaces: The territorialization of resource control. *World Development* 101: 388–399.

Richardson T. and Weszkalnys G. (2014) Introduction: resource materialities. *Anthropological Quarterly* 87 (1): 5–30.

Rivera Andia J.J. and Vindal Ødegaard C. (2019) Introduction: Indigenous peoples, extractivism, and turbulence in South America. In: Vindal Ødegaard C. and Rivera Andía J.J. (eds.) *Indigenous life projects and extractivism: Approaches to social inequality and difference*. Cham: Springer.

Robertson D. (2006) *Hard as the rock itself: Place and identity in the American mining town*, Boulder: University Press of Colorado.

Rogers D. (2015) Oil and anthropology. *Annual Review of Anthropology* 44: 365–380.

Rubbers B. (2019) Mining towns, enclaves and spaces: A genealogy of worker camps in the Congolese copperbelt. *Geoforum* 98: 88–96.

Ruggiero V. and South N. (2013) Toxic state–corporate crimes, neo-liberalism and green criminology: The hazards and legacies of the oil, chemical and mineral industries. *International Journal for Crime, Justice and Social Democracy* 2 (2): 12–26.

Sandlos J. and Keeling A. 2016) Aboriginal communities, traditional knowledge, and the environmental legacies of extractive development in Canada. *The Extractive Industries and Society* 3 (2): 278–287.

Schritt J. and Behrends A. (2018) "Western" and "Chinese" oil zones: Petro-infrastructures and the emergence of new trans-territorial spaces of order in Niger and Chad. In: Engel U., Boeckler M. and Müller-Mahn D. (eds) *Spatial practices: Territory, border and infrastructure in Africa*. Boston: Brill.

Seitz N., van Engelsdorp D. and Leonhardt S.D. (2019) Conserving bees in destroyed landscapes: The potentials of reclaimed sand mines. *Global Ecology and Conservation* 19: e00642.

Shapavalova-Krout D. (2019) International governance of oil spills from upstream petroleum activities in the Arctic: Response over prevention? *The International Journal of Marine and Coastal Law* 34 (4): 668–697.

Sirina A. (2009) Oil and gas development in Russia and northern indigenous peoples. In: Wilson R.E. (ed.) *Russia and the North*. Ottawa: University of Ottawa Press.

Stammler F. and Ivanova A. (2016) Resources, rights and communities: Extractive mega-projects and local people in the Russian Arctic. *Europe-Asia Studies* 68 (7): 1220–1244.

Tsing A. (2005) *Friction: An ethnography of global connection*, Princeton and Oxford: Princeton University Press.

Tsing A. (2017) The buck, the bull, and the dream of the stag: Some unexpected weeds of the Anthropocene. *Suomen Antropologi: Journal of the Finnish Anthropological Society* 42 (1): 3–21.

Vistnes I. and Nelleman C. (2001) Avoidance of cabins, roads, and power lines by reindeer during calving. *Journal of Wildlife Management* 69 (4): 915–925.

Voyles T.B. (2015) *Wastelanding: Legacies of uranium mining in Navajo Country*, Minneapolis: University of Minnesota Press.

Wagreich M. and Draganits E. (2018) Early mining and smelting lead anomalies in geological archives as potential stratigraphic markers for the base of an early Anthropocene. *Anthropocene Review* 5 (2): 177–201.

Watts M. (2001) Petro-violence: Community, extraction, and political ecology of a mythic community. In: Peluso N.L. and Watts M. (eds) *Violent environments*. Ithaca and London: Cornell University Press, 189–212.

White R. (2013) Resource extraction leaves something behind: Environmental justice and mining. *International Journal for Crime, Justice and Social Democracy* 2 (1): 50–64.

Willow A.J. (2012) *Strong hearts, native lands: The cultural and political landscape of Anishinaabe anti-clearcutting activism*, Albany: State University of New York Press.

Willow A.J. (2014). The new politics of environmental degradation: Un/expected landscapes of disempowerment and vulnerability. *Journal of Political Ecology* 21 (1): 237–257.

Willow A.J. (2017) Cultural cumulative effects: Communicating industrial extraction's true costs. *Anthropology Today* 33 (5): 21–26.

Willow A.J. and Wylie S. (2014) Politics, ecology, and the new anthropology of energy: Exploring the emerging frontiers of hydraulic fracking. *Journal of Political Ecology* 21 (1): 237–257.

Yakovleva N. (2011) Oil pipeline construction in Eastern Siberia: Implications for indigenous people. *Geoforum* 42 (6): 708–719.

Zalik A. (2018) Mining the seabed, enclosing the area: Proprietary knowledge and the geopolitics of the extractive frontier beyond national jurisdiction. *International Social Science Journal* 68 (229–230): 343–359.

5

GOVERNANCE

Jeroen Cuvelier, Sara Geenen and Boris Verbrugge

1 Introduction

In reaction to an explicitly normative conception of governance in donor agendas that push good governance and state reconstruction, recent scholarship has taken a more empirical and bottom-up approach. Olivier de Sardan (2011: 22) defines governance as "any organised method of delivering public or collective services and goods according to specific logics and norms, and to specific forms of authority". It is now common wisdom in legal anthropology that in any society, multiple normative and regulatory orders coexist. Aside from states, actors like traditional and customary institutions, militia groups, religious and caste institutions, NGOs, and private companies all engage in norm production and in the provision of public services. By adhering to these norms and by using these services, individuals simultaneously recognize and legitimize the authority of these institutions.

In this chapter, we review anthropological approaches to mineral resource governance, focusing on what is governed as well as on how this governance takes place, by whom, and for whom. Anthropological approaches to resource governance distinguish themselves from other social-scientific approaches by their holistic nature, and the fact that they show greater sensitivity to the particularities of the socio-cultural contexts (Rosen 2020). Moreover, as will become evident in this chapter, anthropology has greatly enriched the scholarly analysis of resource governance dynamics by bringing the complexities of social relations to the fore, scrutinizing how resource governance is fashioned and affected by a wide variety of social connections and interactions at different but entangled levels and scales (see e.g. Ballard and Banks 2003; Fent 2019; Golub 2014; Tsing 2005). In section 2, we provide a broad overview of the actors involved in mineral resource governance, moving from the transnational, over the national, to the local level, while at the same time highlighting the connections between these different levels. In sections 3

DOI: 10.4324/9781003018018-5

and 4, we take a closer look at the governance of land and labour, as the two key factors of production in mineral extraction. This will provide us with clues about who stands to benefit and lose from prevailing governance arrangements.

2 Governance spaces and actors

The extraction and trade of mineral resources have always been quintessentially global phenomena, and they have long been recognized as such by anthropologists. Way before globalization became a buzzword in the social sciences, the Marxist anthropologist Wolf (1982) already drew attention to the large quantities of gold that were being exported overseas from southern Africa, from as early as the 9th century. Yet the most recent wave of globalization has seen an unprecedented "intensified global interconnectedness" (Inda and Rosaldo 2002: 5), which has made the regulation of the mineral industry increasingly complicated (Eriksen 2014: 86). Held and McGrew nicely summed up the crux of the matter when they wrote that "the globalization of economic activity exceeds the regulatory reach of national governments while, at the same time, existing multilateral institutions of economic governance have limited authority because states, jealously guarding their national sovereignty, refuse to cede them substantial power" (Held and McGrew 2000: 26–27). Anthropologists trying to make sense of the governance of the contemporary mining industry have been faced with the daunting task of having to unravel a Gordian knot composed of multiple levels of governance, multiple governing bodies, and multiple normative orders. As such they have been intrigued and preoccupied with what Tsing (2005: 3) has referred to as "the productive friction of global connections". In what follows we think through the transnational, national, and local levels as a way of structuring our chapter. At the same time, we fully recognize that these scales are deeply entangled.

Transnational governance

Since the late 1990s, the transnational expansion of mining has gone hand in hand with a growing involvement of international financial institutions (IFIs) in the governance of mining. Besides the World Bank, the International Finance Corporation (IFC), and the International Monetary Fund (IMF), regional development banks such as the African Development Bank, the Asian Development Bank, the European Bank for Reconstruction and Development, and the Inter-American Development Bank have been acquiring more and more influence. To give but one recent example: on 12 March 2020, the Inter-American Development Bank announced the approval of a US$78.4 million loan for investments in Ecuador's mining and energy sectors, saying that it would simultaneously contribute to developing policy tools to "boost up the regulatory framework's effectiveness and create a business atmosphere that promotes responsible investments" (IADB 2021).

While IFIs are deeply involved in shaping the geography and governance of mining investments across the globe, economically powerful and resource-seeking nation-states such as China play a somewhat different but equally significant role. In her ethnography of China's influence on Zambia's economic development, Lee has pointed out that Chinese state capital distinguishes itself from global private capital in that it is less obsessed with short-term profit maximization. It follows a broader logic of "encompassing accumulation" centred around securing political influence and accessing resources – especially those that are in short supply in China, such as copper. Compared to global private investors, Chinese state-owned companies have shown a greater sensitivity to what is at stake in Zambian politics, as well as a stronger preparedness to compromise around mining sector governance (Lee 2017: 31–33). Conversely, the sudden and massive influx of private Chinese capital into informal and illegal small-scale mining operations has been seen as further eroding the governance capacity of the state. This has been most extensively documented for the case of Ghana, where at least 30,000 Chinese migrants were involved in *galamsey* (small-scale gold mining) activities between 2008 and 2013. They were able to violate the country's mining regulations with impunity thanks to the protection offered by well-placed Ghanaian politicians, public officials, and chiefs, who in return received large sums of money. Meanwhile, the large-scale destruction of landscapes and pollution of riverbeds led to protests by Ghanaian media and civil society, forcing the government to place a moratorium on Artisanal and Small Scale Gold Mining (ASGM) and deport thousands of Chinese nationals. However, according to Crawford and Botchwey (2020), this only served to conceal the endemic corruption among government officials, politicians, and chiefs that has encouraged and protected informal operations.

The spread of mining activities across the globe has also given rise to a marked transnationalization of protest against mining (Jacka 2018: 65). More and more local activist groups have joined hands with international NGOs in criticizing and campaigning against the negative socio-economic and environmental impacts of large-scale mining (LSM) activities (Kirsch 2014; Li 2015). In response to this, and in an attempt to obtain/keep a "social licence to operate", companies have stepped up their efforts to pursue more social, ethical, and sustainable ways of doing business through myriad initiatives in the fields of corporate social responsibility (CSR), ethical trade, business and human rights, and various networked forms of governance involving corporations, industry associations, NGOs, and third party auditors[1] (Dolan and Rajak 2011). In a context of widespread liberalization and state retreat, which was ideologically inspired by the "Washington Consensus" and actively promoted by international lenders, these actors are seen as filling the governance void left behind by weak and fragile states. At the same time, these networked forms of mineral resource governance can also be seen as part of a post-Cold War effort by the Global North to "stabilize" and to maintain control in the Global South (Duffield 2001).

A good illustration of the highly intricate nature of networked forms of mineral resource governance can be found in the international diamond industry. In her multi-sited ethnography, Siegel (2009) describes the interplay between unofficial norms and conventions (codes of honour, entrance ceremonies, rules for the settlement of disputes, handshakes ...), the laws of diamond producing and importing countries, and an ambitious transnational regulatory framework in the form of the Kimberley Process Certification Scheme (KPCS), which was set up to prevent so-called "conflict diamonds" from entering global markets (see also Van Bockstael and Vlassenroot 2009). A number of studies on the impact of similar initiatives to combat the trade in conflict minerals have shown that when such networked forms of governance touch ground, they form new hybrids that often have adverse impacts on their intended beneficiaries, for instance by excluding small-scale producers from global markets (see Geenen 2015 and Vogel and Raeymaekers 2016 for the case of the DRC).

CSR is another excellent example of networked governance. It has often been analysed as governance that is outsourced to private companies (see Campbell 2012 for Africa; Cheshire 2010 for Australia; Haalboom 2012 for Latin America). As Shever (2018: 4) has remarked, the trend of delegating more and more governance tasks to the private sector can be attributed to the neoliberal belief that corporations are more rational, efficient, and effective than states. Anthropological research has shown that CSR offers several advantages to corporations. Apart from the fact that it has enabled them to change their reputation from selfish profit-seekers to capitalists with a friendly face, it has also put them in a position to subtly extend their authority over the people living in and near their areas of operation. Furthermore, CSR interventions have given them the opportunity to present themselves as effective problem-solvers, without having to account for their own roles in creating many of the problems in the first place (Shever 2018). Authors such as Owen and Kemp (2014), Rajak (2011), and Welker (2014) have stressed that corporations should not be studied as monolithic entities, since company policies are always enacted by individual actors with their own personal interests and objectives, navigating relationships of agreement and disagreement within the company itself. Consequently, CSR interventions should not be interpreted as evidence of the existence of one single company philosophy, but rather as the outcome of processes of negotiation, competition, and struggle within the companies themselves.

State governance

In most countries (the United States being a notable exception) the subsurface and the mineral resources contained in it are owned by the state, which consequently has the right to decide whether and by whom mineral resources are extracted, and how mineral revenues are distributed (Emel and Huber 2008). In the neoliberal era, the state has "rolled back" in favour of private actors, but has simultaneously "rolled out" to protect the interests of global capital (Peck and Tickell 2012). There exists widespread consensus that state mining policies favour large-scale, industrial mining operations, often undertaken by transnational mining corporations and having negative socio-environmental effects on local communities. In many cases this is paired with a marginalization and criminalization of artisanal and small-scale mining (ASM). In Latin America, the dominance of highly liberalized foreign direct investment regimes has given rise to an increase of mining investment by more than 100 per cent between 2002 and 2012. This went hand in hand with increased levels of resistance, for which local communities and civil society reached out to transnational networks (Dougherty 2016). In sub-Saharan Africa, similar developments have been observed in countries like Ghana (Hilson and Potter 2005), Burkina Faso (Luning 2008), and Mozambique (Wiegink 2018). In many of these countries, the expansion of corporate mining went hand in hand with the displacement of artisanal miners, and/or the introduction of new policies that hinder the formalization of ASM.

That being said, it is important to note that the image of neoliberal states coalescing with global mining capital is to a certain extent misleading. In many countries, liberalization has intersected with a trend towards democratization and decentralization. States are often characterized by high degrees of internal pluralism, with different actors and agencies adopting diverging views on mining (Spiegel 2012). In the Philippines, for instance, mining is governed by a wide range of contradictory and overlapping laws and administrative rules that are created by different government agencies (Verbrugge 2015a). In this context, three types of resource struggle have erupted: intra-government conflicts over fiscal-regulatory authority in the mining sector; conflicts between large-scale and small-scale mining over access to mineral-bearing land; and conflicts between tribal groups over rights to and revenues from ancestral land.

Furthermore, host governments and local communities (which are not necessarily united and/or internally coherent) have shown growing frustration over the lack of economic benefits accruing to them. In some countries, this has fed into significant policy shifts. Neo-extractivist states in Latin America have increased state control over the mining sector, coupling this to redistributive politics. As Pellegrini has noted in his assessment of mining policies in Bolivia, the term neo-extractivism is used to refer to "a continuation of the extractivist development model that is accompanied by a range of social policies, themselves financed by the revenues generated by extractive activities" (2018: 139). In countries like Tanzania (Pedersen et al. 2019) and Peru (Cortés-McPherson 2019) new forms of resource nationalism are slowly tilting the regulatory balance in favour of domestic mining interests.

As Gilberthorpe and Rajak argue in a recent review article, anthropology has an important role to play in re-embedding and re-politicizing these struggles, by moving away from technical questions of revenue distribution, towards analysing "the constellation of subaltern and elite agency at work within processes of resource extraction (...) in order to confront the complexities, incompatibilities, and inequities in the exploitation of mineral resources" (2017: 186). Heeding this call, van Teijlingen (2016: 910) analyses the neo-extractivist project in Ecuador as a governmentality project that constructs "new subjectivities along the lines of new categories (that of the 'strategic sectors', the 'national interest', the 'poor', and 'the blessed', the patriots and the anti-patriots, etc.), appealing to a historically shaped 'will to improve'" (see also Vela-Almeida 2018). More broadly, popular movements against extractivism in Latin America have been interpreted not merely as struggles about mineral governance, but also about citizenship, the nation, rights, and identity. Rather than advocating a return to previous modes of sovereign state power, Perreault and Valdivia (2010: 698) argue, these movements "seek to carve out a space for state authority that is more closely aligned with what they see as the interests of the population".

Local governance

As rightly noted by Fisher (2008), mining activities often take place "at the margins" of state power, where governance is the product of an interaction between the state (and its decentralized bodies) and various other actors. With this observation, she essentially enters the domain of legal anthropology, which has analysed this empirical reality as a situation of legal pluralism (Von Benda-Beckmann and Von Benda-Beckmann 2006). Political scientists working on state formation and scholars in peace and conflict studies have introduced the notion of hybrid political orders (Boege et al. 2008). Moving away from the failed states paradigm that centres on how states in the Global South deviate from ideal-typical Weberian states, studies of hybrid political orders focus on the "interactions of traditional, personal, kin-based or clientelistic logics with modern, imported, or rational actor logics" (Hoffmann and Kirk 2013: 18). As we explain in further detail below, other social scientists have followed a broadly similar approach, examining public authority in local political arenas as a hybrid of externally imposed orders and existing local orders.

A number of conceptual lenses have been used to analyse instances of pluralism or hybridity in ASM. Building on the work of Moore (1978), Grätz (2002) conceptualizes gold mining communities in West Africa as "semi-autonomous social fields". He discusses a diversity of rules pertaining to revenue sharing, land access, disputes over theft, cheating, and working times. These norms integrate elements from other normative systems, including host–guest relations and patron–client relations. The authorities that enforce norm compliance vary from shaft owners, over elected leaders, to traditional chiefs. In the case of the Eastern DRC, Geenen (2015) uses Olivier de Sardan's (2011) notions of practical norms to describe how a wide range of mining practices (gaining access to a mining pit, sharing output, being recruited in a mining team, paying unofficial taxes, etc.) are sanctioned by state agents, customary chiefs, miners' cooperatives, militia groups, and pit managers. New norms are the result of a "bricolage" process (Cleaver 2012) in which people consciously or unconsciously draw upon pre-existing norms and practices to shape new institutions. For instance, when militia groups occupying an ASM mine in Eastern DRC installed a system of rent payments, this was shaped by existing customary rules, but also by a fraudulent system that was in place when the mine was still operated by an industrial company. In Burkina Faso, Werthmann (2003) describes the case of the elected leader of a recently established mining camp. She argues that the elected leader can be seen as a typical Big Man, combining different sources of power: that of a politician/entrepreneur and that of a "manager of violence". His leadership is based on formal vote and economic wealth, but also on charisma and violence. Macintyre (2008) has similarly highlighted the persistence of Big Man dynamics in Papua New Guinea, a so-called weak state with a long history of plantation agriculture and resource extraction. She describes how Big Men gain and retain their legitimacy through their capacity to distribute consumer goods to their clients. As such, "drinking and sharing beers with others has been interpreted (…) as an extension or transformation of traditional 'big man' activity into the domain of the capitalist economy" (Macintyre 2008: 189).

Another analytical lens for studying hybrid forms of resource governance in ASM is that of the frontier. Several authors draw on Kopytoff's (1987) classic work on the pre-colonial African frontier as an open space situated in the interstices between dominant political orders and kingdoms, which gives rise to the emergence of new political, social, and cultural formations. Applying this idea to the gold mining economy of Burkina Faso, Werthmann (2012) identifies an artisanal mining frontier that came into being during a drought in the 1980s, and an industrial mining frontier that emerged in the wake of the abovementioned neoliberal policy changes. Côte and Korf (2018) similarly use Kopytoff's insights to shed light on the emergence of mining concessions and associated regulatory arrangements. They problematize Ferguson's (2005) portrayal of mining concessions as spaces beyond the reach of the state, by instead describing these environments as being characterized by an excess rather than a lack of regulatory authority – a phenomenon they describe as "the plurification of regulatory authority".

These different accounts make abundantly clear that the state is not the only actor involved in shaping mineral resource governance. Yet even in areas where the state is purportedly absent, references to state power are often ubiquitous. Within a research tradition that is commonly referred to as the anthropology of the state (Sharma and Gupta 2006), different authors remind us that the idea of the state is a potent instrument to legitimize claims to authority (Hansen and Steputat 2006; Trouillot 2001). This idea of the state may be acted upon through a variety of "state practices" and symbolic references to legality. In their book on the Papua New Guinean state, Bainton and Skrzypek (2021: 19) write that "states reinforce the primacy of their sovereign authority through their exclusive claim over the mineral resources in the land. The state's supreme authority is symbolically upheld through formal legislative frameworks that govern the ownership and extraction of these resources". Both in the Eastern DRC (Geenen 2015) and in the southern Philippines (Verbrugge 2015b), local state agents use their legal command to impose a range of unofficial payments on informal miners and traders.

Instances of hybridity have also been described in the context of LSM. Writing about gold mining in Ghana, Geenen (2016) discerns different types of engagement between companies, communities, and governments, which are characterized by three governance modes: conflict and resistance, expectations, and negotiations. In another analysis of Ghana's mining sector, Luning and Pijpers (2017) coin the concept of "in-depth geopolitics" to describe the interplay between three factors shaping mineral governance in places where ASM and LSM interact, notably the stage of a mining operation, the socio-political context and local histories of extraction, and the characteristics of the "subterranean structure". By paying attention to the dynamics of in-depth geopolitics, the authors criticize Ferguson's (2005) argument about the enclaved nature of LSM, which is sometimes seen as disconnected from local social, political, and economic realities. In Latin America, a lot of anthropological work has detailed the complex interactions between multi-national mining companies and local communities (Gustafsson 2018). The focus of this research is often on social mobilization and resistance against LSM (Kirsch 2014; Li 2015). Several authors, however, have warned against romanticizing resistance or considering communities as monolithic entities. This essentially implies opening the black box of the community by highlighting internal conflicts as well as different attitudes, interests, and alliances in relation to the mining project (Conde and Kallis 2012; Geenen and Verweijen 2017; Haalboom 2012; Kapelus 2002; Walter and Urkidi 2017).

3 Governing mineral land

While most mining laws recognize ASM as a possible mode of mineral production, governments often – wittingly or unwittingly – impose a range of bureaucratic and political barriers that prevent ASM operators from obtaining formal rights to mine. Many ASM operators are therefore left with no choice but to operate "illegally", often inside company concessions (Hilson et al. 2020). Yet on the ground, mineral tenure relations often deviate from official regulations. Following Ribot and Peluso (2003), those actors that wield formal rights are not always able to effectively exercise these rights and to derive benefits from their property. Other actors may have access to mineral-bearing land without holding formal property rights. To strengthen their access claims, these actors may seek authorization from different – and possibly competing – institutions, thereby strengthening these institutions in the process. In this way, struggles over access to land (and other natural resources) become entangled with broader struggles over authority (Sikor and Lund 2009).

Quite a lot of attention has been paid to the changing role of customary authorities, particularly (but not exclusively) in the case of sub-Saharan Africa. In ASM areas, their position as distributor and manager of land entitles them to rents from ASM operators. In LSM sites, this same position allows them to negotiate with companies over questions related to (amongst other things) compensation and resettlement. In some countries, like Papua New Guinea, customary land rights have been enshrined in the Constitution (Bainton and Skrzypek 2021). Yet the role of customary authorities has been transformed under the influence of (post) colonial state building and economic modernization. In some cases, the expansion of mining contributes to the erosion of customary power. In South Africa, Nhlengetwa and Hein (2015: 2) claim that so-called *Zama Zama* miners who illegally operate in (abandoned) underground shafts "work outside of customary structures: they do not share their wealth with the community or give a share of earnings to the local chief, who acts as the custodian of the people". Likewise, in her analysis of the erosion of customary rule in Ghana's Ashanti Region, Coyle (2018: 248) notes that "in the power vacuum and amidst the devastation left behind by the absentee chief, it is the *galamsey*, artisanal miners, who often perform the customary functions of governance".

In other cases, customary authorities are more successful in adjusting to new realities. For instance, Cook (2011) describes how the King of the Royal Bafokeng Nation in South Africa's Northwestern province succeeds in being "at once a symbol of traditional authority and a corporate CEO" (Cook 2011: 156). There are also cases where customary authorities' involvement in the mining arena threatens their legitimacy and authority. In West Africa, Lanzano (2018) notes that while customary authorities previously played a central role in regulating access to mining land, they have increasingly turned to rent-seeking and informal taxation. In Tanzania, companies rely on "unaccountable elites for selecting and imple-menting [CSR] projects, resulting in funds being misappropriated to serve private ends" (Lange and Kolstad 2012: 141). Beyond the African context, Golub (2007) and Gilberthorpe and Banks (2012) note how, in Papua New Guinea's LSM areas, "the massive in-flow of cash from resource-derived benefit streams into relatively stable and, importantly, sustainable socio-economic systems" generated inequality, fragmentation and insecurity (Gilberthorpe and Banks 2012: 186). Traditional forms of social organization were replaced by more formalized institutional arrangements from which elites benefited most, and cash inflows related to com-pensation and royalties instigated a culture of individual wealth-seeking. In the DRC (Geenen 2015; Smith 2018) and the southern Philippines (Verbrugge 2015a), the heightened stakes around LSM concessions even led to conflicts between customary factions. Beyond the role of customary authorities, several authors zoom in on the strategies that different actors deploy to obtain access to mineral-bearing land, and to have their access claims recognized by different (and possibly competing) authorities. In her work on the Eastern DRC, Geenen (2015) describes how ASM operators (and traders) negotiate with customary chiefs as well as landowners, government agents, militia groups, and multinational companies. But she also documents how powerful coalitions of local/national elites and cor-porations use mechanisms of access control such as privatization and formalization to displace ASM miners from the land they were working on. In one case where land disputes had been ongoing for several decades, landowners have (mis)used the formalization process to create a new cooperative and to obtain a new permit. While this permit gave them a semblance of state-sanctioned legality, in practice they installed a coercive and violent regime.

Comparing the interaction between ASM and surface land tenure arrangements in the Philippines, Liberia, and the DRC, Verbrugge et al. (2015) note how informal ASM operators often develop mutually beneficial relations with local landowners. Despite pervasive informality, people make constant references to the idea of legality to legitimize their access claims. Authorities in all three countries have seized upon this opportunity – albeit to varying extents – by initiating a process of "informal formalization", relying on the use of official documents and taxes to legitimize the presence of informal ASM. In doing so, they essentially contribute to projecting state rule into the informal mining economy. A similar process has been described in Colombia by Jonkman (2019), where ASM operators engage in "bottom-up law making" by engaging with a variety of state procedures and documents to legitimize their presence.

To conclude this section, we want to draw attention to the question that should be key to analysing such governance arrangements, namely who stands to benefit and lose. For instance, Verbrugge et al. (2015: 58) note that "those who command vital financial resources and political connections, and who have knowledge of bureaucratic procedures, are better equipped to position themselves within the state arena and to use (references to) legality as an access mechanism than those bereft of such resources". More broadly, while hybrid governance arrangements and plurality may offer a lot of room for manoeuvre for example for informal ASM operators who want to have their access claims authorized, they also tend to be highly unstable and prone to elite capture.

4 Governing mining labour

A lot of the earlier anthropological work has focused on labour migration (Harries 1994; Read 1942; Richards 1932; Schapera 1947) and corporate policies to "stabilize" and control the workforce – particularly in the case of Southern Africa (Jeeves 1985; Moodie 1994; Van Onselen 2001). While early 20th-century migrant labour worked on short-term contracts and maintained strong links with rural homesteads, the worker's image that was cultivated from the 1930s onwards would be one of a modern industrial man (Cuvelier 2011; Powdermaker 1962). In colonial Africa, new ideas about modernity, social and family life, and urban–rural relations became inscribed in a broader governmentality project aimed at controlling colonial subjects. In Latin America, miners' labour unions have come to play an important role from the mid-20th century, in countries like Bolivia even substantially weighing on national politics. Anthropologists such as Nash (1979) and Taussig (1980) have provided fascinating accounts of the socio-cultural characteristics of Latin American mining communities, including the role of religion and rituals and the development of class identities. Contemporary anthropological work such as Smith Rolston's (2014) on coal mining in the American West, zooms in on how miners work and live together, and, faced with the challenges and stress in their jobs, "construct kinship" and defy traditional gender roles. The work of Pugliese (2020) on the other hand shows that corporate policies to promote gender equality collide with prevailing gender ideologies.

With globalization and the move away from vertically integrated companies towards more network-like structures, mining companies have also made changes in how they govern labour. The contemporary labour market for LSM in low- and middle-income countries is typically segmented between high-skilled and well-paid jobs for permanent workers; and low-skilled, low-paid, and casual jobs for subcontracted workers (Rubbers 2019). Access to these jobs is mediated through brokers, who are often elite members. They may be owners or shareholders of labour hire companies, or act as recruiters for transnational companies. In his work on Uzbekistan, Markowitz (2008) details several rent-seeking strategies adopted by such "local elites". In countries where LSM activities are displacing ASM, local elites may find in this brokerage role a way to maintain – albeit in a different form – some of the rents as well as the authority they lost when losing control over land and ASM (Geenen 2019). For mining companies, outsourcing to local firms is not only cost-reducing; it also allows them to comply with local content requirements imposed by the government, to win legitimacy at the local level, and to offer local elites something in return for previous rent-seeking possibilities. As Billo (2015) shows for the case of Ecuador, such patronage practices are rooted in long-standing histories of paternalism and dependence. Yet on the part of the company, she argues, they reveal a short-term vision, a lack of care about social welfare, and an unwillingness to take responsibility for the governance of CSR programmes. Moreover, the spatial segmentation of labourers that accompanies their social segmentation allows for companies and states to better control labour and to restrain collective action, as Manky (2018) demonstrates in the case of Peru.

In ASM, labour may be organized in different ways, depending on historical practices, local geologies, and the type of mining operations (Cleary 1990; Verbrugge and Geenen 2020). In truly small-scale operations, miners may work individually or in small teams that are held together by family-, kinship-, or friendship ties. Larger and externally financed operations often have more pronounced labour hierarchies and tend to rely more heavily on migrant labour. Particularly in this latter type of operation, financiers often use subtle and less subtle ways to control workers (Verbrugge et al. 2015). While ASM is often depicted as an open-access activity par excellence, available evidence suggests that social networks, as well as gender and age, play a vital role in regulating access to labour opportunities (Jønsson and Bryceson 2009). Work on Latin America (De Theije 2017), West Africa (Bolay 2014; Lanzano and Arnaldi di Balme 2017) and Southern Africa (Mkodzongi and Spiegel 2020) has shown how miners' incredible mobility is also key to shaping labour organization. Finally, in several countries, mining cooperatives have emerged as important actors in organizing and controlling mining labour in ASM. Evidence from countries as diverse as Bolivia (Salman 2016), the Philippines (Verbrugge and Besmanos 2016) and the DRC (De Haan and Geenen 2016) indicates that these cooperatives often fall prey to elite capture, thereby putting further downward pressures on workers' shares.

5 Conclusion

This chapter has revealed an ongoing process of hybridization, whereby distinctions between formal and informal, modern and traditional, state and non-state, and local, national, and transnational governance are becoming increasingly difficult to maintain. While this process of hybridization creates opportunities for a re-negotiation of access to land, and arguably to labour opportunities, there is ample evidence that it may also contribute to processes of elite capture. Let us end with three suggestions for a future research agenda.

First, there is still ample room for a more intimate engagement with anthropological debates on governance and the state, which could provide conceptual and analytical tools for more systematic comparative analysis. There is also room for more anthropological work on official state bodies – at the national and decentralized level – and their engagement in mineral resources governance (as in recent work by Bainton and Skrzypek 2021).

Second, while this chapter has shed some light on who stands (not) to benefit from the particular ways in which access to land and (to a lesser extent) labour is governed, far less attention has been paid to whether and how ordinary people can exercise agency in an attempt to (re)shape governance arrangements to their own advantage.

Third, existing research has implicitly or explicitly assumed that questions surrounding access to mineral-bearing land are key to understanding the dynamics and outcomes of mineral resource governance. Relatively less attention has been paid to the governance of labour markets, although this figured prominently in the earlier work of anthropologists who studied colonial and postcolonial mining regimes. Recent scholarship has again started to address this issue, and we would encourage anthropologists to pursue this attention to (the governance of) labour.

Note

1   Important examples in the case of mining include the Responsible Minerals Initiative, the Extractive Industries Transparency Initiative, the Kimberley Process, Fairtrade Gold, the OECD Due Diligence Guidance for Responsible Supply Chains, and the EU Conflict Minerals Legislation.

Bibliography

Bainton N. and Skrzypek E. (2021) *An absent presence: Encountering the state through natural resource extraction in Papua New Guinea and Australia*, Canberra: ANU Press.
Ballard C. and Banks G. (2003) The anthropology of mining. *Annual Review of Anthropology* 32: 287–313.
Billo E. (2015) Sovereignty and subterranean resources: An institutional ethnography of Repsol's corporate social responsibility programs in Ecuador. *Geoforum* 59: 268–277.
Boege V., Brown A., Clements K. and Nolan A. (2008) *On hybrid political orders and emerging states: State formation in the context of fragility*, Berlin: Berghof Research Center for Constructive Conflict Management.
Bolay M. (2014) When miners become "foreigners": Competing categorizations within gold mining spaces in Guinea. *Resources Policy* 40: 117–127.
Campbell B. (2012) CSR and development in Africa: Redefining the roles and responsibilities of public and private actors in the mining sector. *Resources Policy* 37: 138–143.
Cheshire L. (2010) A corporate responsibility? The constitution of fly-in, fly-out mining companies as governance partners in remote, mine-affected localities. *Journal of Rural Studies* 26: 12–20.

Cleary D. (1990) *Anatomy of the Amazon gold rush*, Berlin: Springer.

Cleaver F. (2012) *Development through bricolage. Rethinking institutions for natural resource management*, London: Routledge.

Conde M. and Kallis G. (2012) The global uranium rush and its Africa frontier. Effects, reactions and social movements in Namibia. *Global Environmental Change* 22: 596–610.

Cook S.E. (2011) The business of being Bafokeng. The corporatization of a tribal authority in South Africa. *Current Anthropology* 52 (3): 151–159.

Cortés-McPherson D. (2019) Expansion of small-scale gold mining in Madre De Dios: "Capital interests" and the emergence of a new elite of entrepreneurs in the Peruvian Amazon. *The Extractive Industries and Society* 6 (2): 382–389.

Côte M. and Korf B. (2018) Making concessions: Extractive enclaves, entangled capitalism and regulative pluralism at the gold mining frontier in Burkina Faso. *World Development* 101: 466–476.

Coyle L. (2018) Fallen chiefs and sacrificial mining in Ghana. In: Comaroff J., and Comaroff J. (eds) *The politics of custom. Chiefship, capital and the state in contemporary Africa*. Chicago and London: University of Chicago Press, 247–278.

Crawford G. and Botchwey G. (2020) Ghana: Controversy, corruption, and Chinese miners. In: Verbrugge B. and Geenen S. (eds) *Global gold production touching ground: Expansion, informalization, and technological innovation*. London: Palgrave Macmillan, 185–206.

Cuvelier J. (2011) *Men, mines and masculinities: The lives and practices of artisanal miners in Lwambo (Katanga province, DR Congo)*. PhD Thesis, KU Leuven, Belgium.

De Haan J. and Geenen S. (2016) Mining cooperatives in eastern DRC. The interplay between historical power relations and formal institutions. *Extractive Industries and Society* 3 (3): 823–831.

De Theije M. (2017) Small-scale gold mining in the Guianas. Mobility and policy across national borders. In: Hoefte R., Bishop M. and Clegg P. (eds) *Post-colonial trajectories in the Caribbean: The three Guianas*. New York: Routledge, 92–106.

Dolan C. and Rajak D. (2011) Introduction: Ethnographies of corporate ethicizing. *Focaal – Journal of Global and Historical Anthropology* 60: 3–8.

Dougherty M. (2016) From global peripheries to the earth's core. The new extraction in Latin America. In: Deonandan K. and Dougherty M. (eds) *Mining in Latin America: Critical approaches to the new extraction*. Abingdon: Routledge, 3–24.

Duffield M. (2001) *Global governance and the new wars*, London: Zed Books.

Emel J. and Huber M.T. (2008) A risky business: Mining, rent and the neoliberalization of "risk". *Geoforum* 39: 1393–1407.

Eriksen T.H. (2014) *Globalization: The key concepts* (2nd edition). London, New Delhi, New York and Sydney: Bloomsbury.

Fent A. (2019) Governing alongside: Lateral spatiality and unmet expectations amid mining negotiations in Casamance, Senegal. *American Ethnologist* 46 (1): 20–33.

Ferguson J. (2005) Seeing like an oil company: Space, security, and global capital in neoliberal Africa. *American Anthropologist* 107 (3): 377–382.

Fisher E. (2008) Artisanal gold mining at the margins of mineral resource governance: A case from Tanzania. *Development Southern Africa* 25 (2): 199–213.

Geenen S. (2015) *African artisanal mining from the inside out. Access, norms and power in Congo's gold sector*, Abingdon: Routledge.

Geenen S. (2016) *Hybrid governance in mining concessions in Ghana*. IOB Working Paper 2016.05. Antwerp: University of Antwerp.

Geenen S. (2019) Gold and godfathers: Local content, politics, and capitalism in extractive industries. *World Development* 123: 104605.

Geenen S. and Verweijen J. (2017) Explaining fragmented and fluid mobilization in gold mining concessions in eastern Democratic Republic of the Congo. *The Extractive Industries and Society* 4 (4): 758–765.

Gilberthorpe E. and Banks G. (2012) Development on whose terms?: CSR discourse and social realities in Papua New Guinea's extractive industries sector. *Resources Policy* 37: 185–193.

Gilberthorpe E. and Rajak D. (2017) The anthropology of extraction: Critical rerspectives on the resource curse. *The Journal of Development Studies* 53 (2): 186–204.

Golub A. (2007) Ironies of organization: Landowners, land registration, and Papua New Guinea's mining and petroleum industry. *Human Organization* 66 (1): 38–48.

Golub A. (2014) *Leviathans at the gold mine: Creating indigenous and corporate actors in Papua New Guinea*, Durham: Duke University Press.

Grätz T. (2002) *Gold mining communities in Northern Benin as semi-autonomous social fields*. Working Paper 36. Halle, Germany: Max Planck Institute for Social Anthropology.

Gustafsson M.T. (2018) *Private politics and peasant mobilization*, London: Palgrave Macmillan.

Haalboom B. (2012) The intersection of corporate social responsibility guidelines and Indigenous rights: Examining neoliberal governance of a proposed mining project in Suriname. *Geoforum* 43: 969–979.

Hansen T.B. and Steputat F. (2006) Sovereignty revisited. *Annual Review of Anthropology* 35: 295–315.

Harries P. (1994) *Work, culture and identity: Migrant labourers in Mozambique and South Africa, c. 1860–1910*, London: James Currey.

Held D. and McGrew A. (2000) The great globalization debate: An introduction. In: Held D. and McGrew A. (eds) *The global transformations reader*. Cambridge: Polity, 1–50.

Hilson G. and Potter C. (2005) Structural adjustment and subsistence industry: Artisanal gold mining in Ghana. *Development and Change* 36 (1): 103–131.

Hilson G., Sauerwein T. and Owen J. (2020) Large and artisanal scale mine development: The case for autonomous coexistence. *World Development* 130: 104919.

Hoffmann K. and Kirk T. (2013) *Public authority and the provision of public goods in conflict-affected and transitioning regions*. The Justice and Security Research Programme Paper 7. London: LSE.

IADB (2021) *IDB to support sustainable development of Ecuador's mining and energy sectors*. 12 March 2020. Available at: iadb.org (accessed on 19 January 2021).

Inda J.X. and Rosaldo R. (eds) (2002) *The anthropology of globalization: A reader*, Malden: Blackwell Publishers.

Jacka J. (2018) The anthropology of mining: The social and environmental impacts of resource extraction in the mineral age. *Annual Review of Anthropology* 47: 61–77.

Jeeves A.H. (1985) *Migrant labour in South Africa's mining economy: The struggle for the gold mines' labour supply, 1890–1920*, Montreal: McGill-Queen's University Press.

Jonkman J. (2019) A different kind of formal: Bottom-up state-making in small-scale gold mining regions in Chocó, Colombia. *The Extractive Industries and Society* 6 (4): 1184–1194.

Jønsson J.B. and Bryceson D.F. (2009) Rushing for gold: Mobility and small-scale mining in East Africa. *Development and Change* 40 (2): 249–279.

Kapelus P. (2002) Mining, corporate social responsibility and the "community": The case of Rio Tinto, Richards Bay Minerals and the Mbonambi. *Journal of Business Ethics* 39 (3): 275–296.

Kirsch S. (2014) *Mining capitalism: The relationship between corporations and their critics*, Berkeley: University of California Press.

Kopytoff I. (ed.) (1987) *The African frontier: The reproduction of traditional African societies*, Bloomington: Indiana University Press.

Lange S. and Kolstad I. (2012) Corporate community involvement and local institutions: Two case studies from the mining industry in Tanzania. *Journal of African Business* 13 (2): 134–144.

Lanzano C. (2018) Gold digging and the politics of time: Changing timescapes of artisanal mining in West Africa. *The Extractive Industries and Society* 5 (2): 253–259.

Lanzano C. and Arnaldi di Balme L. (2017) "Burkinabe shafts" in Haute Guinée: Technical knowledge circulation in the artisanal mining sector. *Autrepart* 82 (2): 87–108.

Lee C.K. (2017) *The specter of global China. Politics, labor and foreign investment in Africa.* Chicago and London: The University of Chicago Press.

Li F. (2015) *Unearthing conflict. Corporate mining, activism, and expertise in Peru,* Durham: Duke University Press.

Luning S. (2008) Liberalisation of the gold mining sector in Burkina Faso. *Review of African Political Economy* 117: 387–401.

Luning S. and Pijpers R. (2017) Governing access to gold in Ghana: In-depth geopolitics on mining concessions. *Africa* 87: 758–779.

Macintyre M. (2008) Police and thieves, gunmen and drunks: Problems with men and problems with society in Papua New Guinea. *The Australian Journal of Anthropology* 19 (2): 179–193.

Manky O. (2018) Part-time miners. Labor segmentation and collective action in the Peruvian mining industry. *Latin American Perspectives* 45 (5): 120–135.

Markowitz L.P. (2008) Local elites, prokurators and extraction in rural Uzbekistan. *Central Asian Survey* 27 (1): 1–14.

Mkodzongi G. and Spiegel S. (2020) Mobility, temporary migration and changing livelihoods in Zimbabwe's artisanal mining sector. *The Extractive Industries and Society* 7 (3): 994–1001.

Moodie T.B. (1994) *Going for gold. Men, mines and migration,* Berkeley: University of California Press.

Moore S.F. (1978) Law and social change: The semi-autonomous social field as an appropriate subject of study. In: Moore S.F. (ed.) *Law as process.* London: Routledge, 54–81.

Nash J. (1979) *We eat the mines and the mines eat us: Dependency and exploitation in Bolivian tin mines,* New York: Colombia University Press.

Nhlengetwa K. and Hein K. (2015) *Zama-Zama* mining in the Durban Deep/Roodepoort area of Johannesburg, South Africa: An invasive or alternative livelihood? *The Extractive Industries and Society* 2: 1–3.

Olivier de Sardan J.P. (2011) The eight modes of local governance in West-Africa. *IDS Bulletin* 42 (2): 22–31.

Owen J.R. and Kemp D. (2014) Mining and community relations: Mapping the internal dimensions of practice. *The Extractive Industries and Society* 1: 12–19.

Peck J. and Tickell A. (2012) Apparitions of neoliberalism: Revisiting "jungle law breaks out." *Area* 44 (2): 245–249.

Pedersen R.H., Mutagwaba W., Jønsson, J.B., Schoneveld G., Jacob T., Chacha M. et al. (2019) Mining-sector dynamics in an era of resurgent resource nationalism: Changing relations between large-scale mining and artisanal and small-scale mining in Tanzania. *Resources Policy* 62: 339–346.

Pellegrini L. (2018) Imaginaries of development through extraction: The "history of Bolivian petroleum" and the present view of the future. *Geoforum* 90: 130–141.

Perreault, T. and Valdivia, G. (2010) Hydrocarbons, popular protest and national imaginaries: Ecuador and Bolivia in comparative context. *Geoforum* 41: 689–699.

Powdermaker H. (1962) *Copper town: Changing Africa, the human situation on the Rhodesian Copperbelt,* New York: Harper & Row.

Pugliese F. (2020) Mining companies and gender(ed) policies: The women of the Congolese Copperbelt, past and present. *The Extractive Industries and Society*. DOI: doi:10.1016/j. exis.2020.08.006.

Rajak D. (2011) *In good company. An anatomy of corporate social responsibility*, Palo Alto: Stanford University Press.

Read M. (1942) Migrant labour in Africa and its effects on tribal life. *International Labor Review* 45 (6): 605–631.

Ribot J.C. and Peluso N.L. (2003) A theory of access. *Rural Sociology* 68 (2): 153–181.

Richards A. (1932) Anthropological problems in north-eastern Rhodesia. *Africa* 5 (2): 121–144.

Rosen L.C. (2020) *Fires of gold: Law, spirit and sacrificial labor in Ghana*, Berkeley: University of California Press.

Rubbers B. (2019) Mining boom, labour market segmentation and social inequality in the Congolese Copperbelt. *Development and Change* 51: 1555–1578.

Salman T. (2016) The intricacies of "being able to work undisturbed": The organization of alluvial gold mining in Bolivia. *Society and Natural Resources* 29 (9): 1124–1138.

Schapera I. (1947) *Migrant labour and tribal life: A study of conditions in the Bechuana Protectorate*, London: Oxford University Press.

Sharma A. and Gupta A. (2006) *The anthropology of the state. A reader*, Maiden: Blackwell Publishing.

Shever E. (2018) Transnational and multinational corporations. In: Callan H. (ed.) *The international encyclopedia of anthropology*, Wiley online.

Siegel D. (2009) *The mazzel ritual: Culture, customs and crime in the diamond trade*, Dordrecht, Heidelberg, London, and New York: Springer.

Sikor T. and Lund C. (2009) Access and property: A question of power and authority. *Development and Change* 40 (1): 1–22.

Smith J. (2018) Colonizing Banro: Kingship, temporality, and mining of futures in the goldfields of South Kivu, DRC. In: Comaroff J. and Comaroff J. (eds) *The politics of custom. Chiefship, capital and the state in contemporary Africa*. Chicago and London: University of Chicago Press, 279–304.

Smith Rollston J. (2014) *Mining coal and undermining gender: Rhythms of work and family in the American west*, New Brunswick: Rutgers University Press.

Spiegel S.J. (2012) Governance institutions, resource rights regimes, and the informal mining sector: Regulatory complexities in Indonesia. *World development* 40 (1): 189–205.

Taussig M. (1980) *The devil and commodity fetishism in South America*, Chapel Hill: University of North Carolina Press.

Trouillot M. (2001) The anthropology of the state in the age of globalization: Close encounters of the deceptive kind. *Current Anthropology* 42 (1): 125–138.

Tsing A. (2005) *Friction: An ethnography of global connection*, Princeton and Oxford: Princeton University Press.

Van Bockstael S. and Vlassenroot K. (2009) From conflict diamonds to development diamonds: The Kimberley Process and Africa's artisanal diamond mines. *Studia Diplomatica* 62 (2): 79–96.

Van Onselen C. (2001) *New Babylon, new Nineveh: Everyday life on the Witwatersrand 1886–1914*, Johannesburg and Cape Town: Jonathan Ball Publishers.

Van Teijlingen K. (2016) The "will to improve" at the mining frontier: Neo-extractivism, development and governmentality in the Ecuadorian Amazon. *The Extractive Industries and Society* 3: 902–911.

Vela-Almeida D. (2018) Territorial partitions, the production of mining territory and the building of a post-neoliberal and plurinational state in Ecuador. *Political Geography* 62: 126–136.

Verbrugge B. (2015a) Decentralization, institutional ambiguity, and mineral resource conflict in Mindanao, Philippines. *World Development* 67: 449–460.

Verbrugge B. (2015b) Undermining the state? Informal mining and trajectories of state formation in Eastern Mindanao, Philippines. *Critical Asian Studies* 47 (2): 177–199.

Verbrugge B., Cuvelier J. and Van Bockstael S. (2015) Min(d)ing the land: The relationship between artisanal and small-scale mining and surface land arrangements in the Southern Philippines, Eastern DRC and Liberia. *Journal of Rural Studies* 37: 50–60.

Verbrugge B. and Besmanos B. (2016) Formalizing artisanal and small-scale mining: Whither the workforce? *Resources Policy* 47: 134–141.

Verbrugge B. and Geenen S. (2020) *Global gold production touching ground: Expansion, informalization, and technological innovation*. London: Palgrave Macmillan.

Vogel C. and Raeymaekers T. (2016) Terr(it)or(ies) of peace? The Congolese mining frontier and the fight against "conflict minerals". *Antipode* 48 (4): 1102–1121.

Von Benda-Beckmann F. and Von Benda-Beckmann K. (2006) The dynamics of change and continuity in plural legal orders. *Journal of Legal Pluralism and Unofficial Law* 53–54:1–44.

Walter M. and Urkidi L. (2017) Community mining consultations in Latin America (2002–2012): The contested emergence of a hybrid institution for participation. *Geoforum* 84: 265–279.

Welker M. (2014) *Enacting the corporation: An American mining firm in post-authoritarian Indonesia*, Berkeley: University of California Press.

Werthmann K. (2003) The president of the gold diggers. Sources of power in a gold mine in Burkina Faso. *Ethnos* 68 (1): 95–111.

Werthmann K. (2012) Gold mining in Burkina Faso since the 1980s. In: Werthmann K. and Grätz T. (eds) *Mining frontiers in Africa: Anthropological and historical perspectives*. Cologne: Rüdiger Köppe Verlag, 119–132.

Wiegink N. (2018) Imagining booms and busts: Conflicting temporalities and the extraction–"development" nexus in Mozambique. *The Extractive Industries and Society* 5 (2): 245–252.

Wolf E. (1982) *Europe and the people without history*, Berkeley and Los Angeles: University of California Press.

6

MATERIALITY AND SUBSTANCES

Elizabeth Ferry

1 Introduction

Humans have used minerals for millennia, and in most cases, in order to use them, they have to remove them from their original context more or less forcefully, shape them, and sometimes put them to use with other minerals, vegetables, or animals. However, a defined Anthropology of Resource Extraction field is quite new. Attention to resources – "objects and substances extracted from 'nature' for human enrichment and use" (Ferry and Limbert 2008: 4) – and extraction – the process of separating objects and substances from their surroundings and rendering them as resources – has risen strongly in anthropology in the past 20 years. Along with this surge in attention to resource extraction has come an intensified focus on these processes as fundamentally material ones, and an attention to activities and formations of resource extraction as occurring within historically embedded material-social composites.

I take the terms "materials", "materiality",[1] and "the material world" as covering more or less the same conceptual terrain, referring to what Douglas Rogers calls "the sensuous and phenomenal qualities of things and their implication in human social and cultural life" (2012: 285). An attention to materiality would therefore mean including things and their qualities not as mere background but as integral dimensions of an analysis. This definition can encompass a range of recent approaches to the material world.[2]

I also use the term "substance" to refer to some extracted materials. The term emphasizes the sensuous properties of materials and their seeming categorical status (oil as a category of material, rather than an individual object). A substance is a kind of assemblage, to be sure, but a homogenous one (or one that is perceived to be so). This concept of a relatively undifferentiated assemblage encompasses many of the materials that emerge in extraction and that are covered in this chapter, such as:

DOI: 10.4324/9781003018018-6

gemstones, bauxite, silver, gold, copper, etc. It helps to us to consider (real and perceived) homogeneity as a materially and semiotically meaningful characteristic of some extracted materials, and therefore to explore the physical properties that come to play such a large semiotic and affective role in the lives of these substances and the people and things with which they come into contact.

After discussing how (a few) anthropologists working on resource extraction prior to the 2000s have mobilized, discussed, and theorized the material world, this chapter will turn to a thematic consideration of more recent works on mining, in which materials are more frontally present and more deeply theorized. The chapter concludes with a brief discussion of emergent and new areas for inquiry.

2 Materiality in early anthropologies of extraction

In the 20th century, anthropological case studies of mining did not coalesce in a distinctive subfield with its own terrain, questions, and conversations. Instead, these studies tended to contribute to other debates or schools of thought such as political economy (Nash 1993), symbolic anthropology (Biersack 1999; Jorgensen 1998), the anthropology of migration (Epstein 1958; Gluckman 1944; Powdermaker 1962), of development (Ferguson 2005), of money (Harris, 1989) and, by the turn of the, 21st century, of resistance and social movements (Kirsch 2002; Sawyer 2004). In their contributions to these varied discussions, anthropologists of extraction did not often focus on the actual materials and substances being extracted as commodities, byproducts, or waste, or on the bodies of miners as such. Nevertheless, a few studies did engage with the materials of mining more thoroughly.

Olivia Harris's work on money in Oruro, Bolivia, based on fieldwork from the 1970s, demonstrates an intense and rich engagement with materials. She describes the various ways in which silver, coins, pyrite, and other materials are mobilized in ritual exchanges that bind people into close connection with the mountain. In this context, the products of the mines are seen as created by devils (indexed by the Tío who sits at the entrance to most mines) but nevertheless useful to humans. Harris writes that:

> Money [in the form of ore] comes from the same sources that ensure the potato and maize harvest, and the reproduction of the flocks. By offering food, humans enter into communication with this source – the shadowy domain of the devils or *saxra*. However, money [in the form of coins or worked metals] also comes from the state and corresponds to the sphere of good government, God and the Catholic Church.
>
> *(Harris 1989: 254–255)*

This crucial distinction between metal in the form of chunks of ore and metal that is "*sellado*" (stamped) makes silver's malleability into a material idiom for the cosmic division between the underground realm of the devils and the surface world of the Church.

In an article focused on an explosion in the 1990s of small-scale gold mining in the Mt. Kare area just south of the fabulously productive Porgera gold mine in Papua New Guinea, Aletta Biersack examines the symbolic constructions of the material by the Paiela people who live in the area. The gold, discovered in the late 1980s to be present in the rivers and sands of Mt. Kare, is thought to be the flesh of a Paiela ancestor who takes the form of a gigantic python. "When the python's body exploded; his skin scattered; the gold arose; the land, with all its people and its resources, originated. Although the gold is primeval, only now has it been discovered. The python's secret is revealed" (1999: 78). She describes this material connection between gold and serpent's flesh as a "hermeneutics of the concrete ... focused on organic life and form, on moral and regenerative bodies of different sorts" (p. 70). Biersack shows how gold's material substance, as part of the regenerative body of the python, demonstrates to Paiela people that it is an ancestral, though until recently unknown, bequest.[3]

Although materials are not an explicit focus of June Nash's ethnography *We Eat the Mines and the Mines Eat Us*, her Marxian materialism and her superb ethnographic skills work together to bring out the ways in which material objects, including mined substances, play fundamental roles in constructing Andean lived worlds. In particular, her descriptions of the mines as a world of darkness, smoke, noise, moisture, extremes of temperature, heights and drops evokes the materiality of mining in ways that few mining ethnographies of the time (and thereafter) have done. She mobilizes her interlocutors' metaphor of the mine as a living body, saying, for instance "The caverns of the mine are like a living organism, draining fluids from the wounds inflicted by the workers" (1993: 179). Furthermore, in her descriptions of the *k'araku* or sacrifice of llamas to Supay, the supernatural power of the mountains and underground, and other rituals (chapter 5) Nash does not primarily focus on the substances that are exchanged – rocks, llamas' blood, alcohol, confetti and paper streamers, etc. – but her meticulous ethnography shows the central role these exchanges play in the experience of work and worker solidarity.

To a lesser extent, Moodie and Ndatshe's ethnography of social, political and sexual forms of solidarity and hierarchy in South Africa foregrounds a view of the social and political as paramount and independent categories with the materials of mining as either background or exemplification, in ways that are characteristic of much work on mining. Yet their descriptions of the physical descriptions of work and danger as bases for solidarity, such as the development of "tacit skills...to detect imminent collapses from the behavior of the rock" (1992: 17) and of the importance of physical strength and "fierce masculine pride in holding down such a tough job (p. 18) glance at the material world of deep-pit gold mining.

These four studies distinguished themselves from most other ethnographies of mining prior to 2000 with their attention to the material circumstances and experiences of mining, seen in some cases (Biersack; Harris) through a lens of cosmology and in others (Nash; Moodie and Ndatshe) through one of labour.

3 Material turns, the anthropology of resource extraction in the 21st century

Theoretical and methodological engagements with the material have become widespread within social theory over the past 20 years, and these engagements have also flourished in the anthropology of resource extraction (Akong 2020). These studies rest on an assumption that, as Tanya Richardson and Gisa Weszkalnys put it "natural resources are inherently distributed things whose essence or character is to be located neither exclusively in their biophysical properties nor in webs of socio-cultural meaning" (2014: 8). Recent work has ramified these non-dichotomous approaches to the material and meaningful in resource extraction, along various paths.

Studies published since about 2005 – typically based on ethnographic fieldwork done in the years of a global commodities boom between approximately 2003 and 2014 – have expanded the field considerably, amounting almost to a "boom" in their own right. In what follows I organize the anthropological scholarship on extraction since 2005 into a series of themes, intended to highlight a number of important things that attention to the material world can show us about extraction. The themes are: qualities and affordances; materials in counterpoint; infrastructures; waste; bodies and substance; and values and valuation. I conclude with a brief discussion of emergent and future directions.

Qualities and affordances

The shift in social theory towards taking things as actors or actants with their own irreducible impingements has opened the way for new analyses of the properties of objects as material "affordances" (Gibson 2014; Ingold 2012, 2018; Keane 2018) that create worlds in interaction with each other, including with people. In the realm of extractive resources, this has opened up a wealth of analysis, much of which is focused on the properties and affordances of substances. For instance, Mimi Sheller, in her book *Aluminum Dreams: The Making of Light Modernity* (2014) describes lightness, pliability, and lustre as integral dimensions of the worlds that aluminium produces, both in its production and its use in iconic objects of modernity such as skyscrapers, aeroplanes, and mobile homes. Nusrat Chowdhury (2016) notes how coal in Pulbhari, northwestern Bangladesh, yet to be mined, is described by interlocutors as shiny or glittery, using the same term used for gold or precious gems (*chokchok kora*) and that this putative quality stands in for wealth, goodness, and abundance. David Kneas (2018) examines how geological mapping and modelling projects that posit copper mineralization in Ecuador draw both on the characteristics of porphyry copper deposits and perceptions of a homogenous Andean nature. Kneas' article posits that estimates of resource (especially copper) wealth in Ecuador depend on the priority of certain perceptions of Andean nature over others – ideas of uniformity and equivalence over Andean difference and hetereogeneity. His argument exemplifies how attention to homogeneity and extension of Andean plate tectonics *as material qualities* play a central role in resource-making.

Ethnographers of mining in vastly different contexts have noted the ways in which mined substances and their attendant qualities and affordances can be associated with different genders or seen as themselves gendered. Daniele Moretti describes how ASM gold mining in Mount Kainda, PNG is generally restricted to men, as a result both of so-called "traditional" and colonial divisions of labour, and the idea that "gold dreams", which are instilled by the spirits of the mine and which help instruct potential miners, tend to come only to men, and that the spirits are often thought to be female, although she also found evidence of the reverse (Moretti 2006). In the mines of Oruro, Bolivia and Guanajuato, Mexico it is the veins of tin and silver themselves which are thought to be gendered as female, and therefore off limits to women (Ferry 2005; Nash 1993). In West Africa, mining work below the surface tends to be associated with men, while surface mining and other activities like washing gold are more closely tied to women. At the same time, Pijpers notes that in Pumduru, Sierra Leone, where both gold and diamond mining are substantial sources of livelihood, women are associated with and more involved with gold mining and men with diamond mining, thus demonstrating the wide prevalence of gendered associations of mining and also the highly specific and contextual nature of those associations (Pijpers 2011: 1075).

Some of these scholarly engagements also fold in an attention to the ways that material properties can be mobilized ideologically, especially so as to "naturalize" particular regimes of value and arrangements of power, through a process of fetishism, or as Richard Parmentier has put it, "the naturalization of convention" (1994). For instance, in his meditation on gold and cocaine in Colombia, *My Cocaine Museum*, Michael Taussig engages with gold and cocaine as substances, stating that "it is precisely as mineral and vegetable matter that [gold and cocaine] appear to speak for themselves and carry the weight of human history in the guise of natural history" (2009: xviiii).

The semiotic concept of "qualia" (phenomenal qualities indexically linked to objects that come to operate iconically [Chumley and Harkness 2013]) has provided a helpful framework for some anthropologists to theorize the substances of abstraction. For instance, Filipe Calvão's (2013) work on the clarity and translucence of diamonds argues that these operate as qualia of speed, invisibility, and value. He thus provides a materially oriented perspective on processes of transparency and the ever-present issue of origins in the diamond trade.

Gold is another substance in which material properties play an especially forceful role and one amenable to the analysis of qualic force. Ferry (2016) describes ways in which gold's material properties of mass, lustre, malleability, and non-reactivity (so that it does not tarnish or rust) underscore arguments of its intrinsic, immediate value. That is, gold's properties operate semiotically to create the claim that gold is not a sign of value, but value itself.

The direction and logics of these mobilizations of gold's qualities is not a foregone conclusion, however, in spite of what some observers, including those most attached to gold, might suggest. Arguing against a Eurocentric narrative of gold as inevitably valuable, Les Field (2012) traces how the metaphysics of gold in Precolumbian and European contexts animate its physical characteristics, but in markedly different ways.

Thomas Abercrombie also takes a semiotic approach, noting that "miners [in Potosí, Bolivia] are ... participants in the 'spiritual' economy of distributed agency that locally sustains the production and movement of minerals and money" (2016: 83). One kind of thing that circulates within this spiritual economy is chunks of ore that, in Abercrombie's terms:

> iconically resemble and indexically point to larger loads of ore, and sums of money, that might be taken from the mountain. ... Gifting and exchanging these chunks of ore ... indexes the value of the social ties that they thereby create and nurture.
>
> *(99–100)*

In their edited volume *Subterranean Estates: Life Worlds of Oil and Gas* (2015), Appel et al. invoke a notion of "petrolic semiosis" (p. 14), and this notion captures a number of works within anthropology that describe how oil's properties have been activated. In particular, oil's blackness (Apter 2008), depth (Rogers 2012), toxicity (Sawyer 2004), and viscosity (Hitchcock 2015) have all figured as material manifestations of its "mythic" (Adunbi 2013) and "magical" (Coronil 1997) power.[4]

These authors, who have different ways of theorizing the relationship between materials and semiosis, all put properties and affordances at the centre of their analyses, something done only glancingly in prior work on extraction.

Materials in counterpoint

A number of authors have noted how materials operate in counterpoint with each other, both in material processes of extraction, distribution, and evaluation, and discursively. Their works recognize that the substances of mining and extraction come out of their matrices and move through the world in interaction with other substances, which help to make them, for instance, valuable commodities, byproducts, and waste, or which define them in opposition or collaboration with other materials used in livelihoods alongside extraction (which may be emergent, displaced, or coincident).[5]

In addition, while all extracted substances that become commodities, by definition, are turned into money and then into other things, gold and silver's histories of use as money makes currencies (dollars, Euros, Bitcoin) a prominent counterpoint in many contexts (Ferry 2016; Maurer et al. 2013). Oil, on the other hand, has played such a central role in the structuring of contemporary financial worlds and regional economies as to appear frequently in other forms like "petrodollars" (Appel et al. 2015; Coronil 1997), or as in Oman's "dreamtime" of oil prosperity, as luxury cars, national schools, and gleaming, up-to-date infrastructure (Limbert 2010).

Other works focus on the counterpoint between materials as part of a political and moral field in which extracted materials are contrasted with other types of thing, or other types of being, often incommensurate to them. Veronica Davidov (2014) discusses how land, copper, and flora comprise competing "metonymic materialities" that stand in for the Intag region of Ecuador, so that:

> As if in a game of "rock, paper, scissors," geological surveys commissioned by the World Bank map that locate profitable subsoil minerals in the region are countered by biome maps, biodiversity surveys, and conservation plans. Quite literally, the emergence of such dominant materialities maps the region, as fertile soil, subterranean minerals, or clusters of specific plants or mosses claim or vie for dominant symbolic status, strategically "deployed" by the state, transnational corporations, and local activists alike.
>
> *(p. 38)*

Mandana Limbert (2016) explores the ways in which water and oil are contrasted in Omani historiography, asking "How do different natural resources, therefore, figure in temporal imaginings? How do they frame the relationships and chronologies of transformation and, more particularly, of causality?" Whereas oil appears only scantly in textbooks and official histories of the Omani sultanate, and appears associated more with the uncertain future than the past, water (particularly in the form of rain) is seen as the market of divine presence and beneficence, and stories of the discovery of particular sources of water are carefully kept and retold. This counterpoint between oil and water provides a material idiom for contrasting temporalities and theories of divine and human causality in Oman.

Infrastructures

Another key point of entry for attention to materials and substance is through the infrastructure of extraction. The social science literature on extraction is by now enormous, and beyond the scope of this chapter, but its premises have had specific effects on how substances and materiality are approached in the anthropology of extraction, and it is from this angle that I introduce it here. Larkin defines infrastructure as "built networks that facilitate the flow of goods, people, or ideas and allow for their exchange over space." (2013: 328) and attention to these built networks has provided a distinctive and useful perspective on 21st century studies of extraction.

The infrastructures of oil have generated particular interest, inspired by several foundational pieces. In his book *Carbon Democracy* (2011), Timothy Mitchell addresses not only the relationship between oil and democracy, but oil *as* democracy, noting that most discussions of oil "have little to do with the ways oil is extracted, processed, shipped and consumed" (p. 2). By including attention to oil's infrastructure of production, financialization, transport, and consumption, as material entities and processes, Mitchell shows the emergence of a growth model of development and militarized control of democratic politics in ways we would not otherwise see. In this way, Mitchell's argument, in addition to what it tells about a host of issues we thought we already understood (like the "resource curse" and "Jihad vs. McWorld"), gives a demonstration of how a material perspective on infrastructure makes a difference.

Michael Watts' formulation of oil assemblages (Watts 2012) forms the basis of a number of generative studies of oil infrastructure. For example, in her discussion of offshore oil production, Hannah Appel attends to the mobile and modular infrastructures that help to bring the "offshore" into being and practice and to contribute to a goal of "frictionless profit" through mobile technologies, infra-structures, workforces, and risk-avoidance regimes in the middle of the ocean that frame the industry's work as self-contained, separate from the local condi-tions in which they are, in fact, so deeply implicated and on which they rely. (Appel 2012: 706). This attention to the sociomaterial infrastructures and proto-cols of oil rigs off Equatorial Guinea shows the "thickness" of what goes into making production socially "thin". Correspondingly, Douglas Rogers looks at how natural gas pipelines in the Perm region of Russia provoked a sense that those who managed them were parasitically profiting from "the material qualities of gas as it moved through this pipeline infrastructure" in particular its trans-portability in contrast to the depths and stability of subsoil deposits (Rogers 2012: 290).

Some studies engage questions of the different socialities that emerge in relation to different infrastructures. One of the most influential of these has been James Ferguson's schematic of "thick" and "thin" sociality, which divides extractive industries according to the degree and type of infrastructure and institutions (2005). For instance, Fabiana Li's (2015) comparison of the smelting complex of la Oroya and the proposed Conga open-pit gold mine exemplifies the distinction between these types of extraction and accompanying socialities.

A focus on transport and its infrastructural and material constraints and affor-dances motivates other scholars. In his book *Material Politics: Disputes along the Pipeline* (2013), Andrew Barry traces the control and mapping of oil as a substance in transport that produces and becomes entangled with knowledge and con-troversy. He argues for an inclusion of material qualities, design, and infrastructure as neither background to nor useful examples of knowledge and politics, but con-stitutive of these.

In his 2008 piece on "improvisational economies" in coltan production in DRC, Jeffrey Mantz elucidates another dimension of the infrastructure of resource extraction; he shows how global digital infrastructures of connectivity constitute and are also constituted by various forms of violence, such as the funding of war-lords with currency obtained by selling coltan, the fantasy violence of Halo3 and other Playstation games made possible with coltan, and shootings at Best Buy stores in the 2006 Christmas season over scarce PS3 units. Mantz takes the infrastructure of digital modernity and of extraction as two aspects of coltan production that exemplify the predicament of global connectedness, stating that the entire promise of an interconnected world arguably has as its precondition this commodity and its particular qualities (of density and relative accessibility, for example), as well as the labour relations, trade conditions, and internal fragmentation that have made [it] available to the world.

(Mantz 2008: 48)

Stefanie Graeter (2020) also explores infrastructure as a technology of violence in her discussion of the "infrastructural incorporations" of lead accumulated in the bodies of residents of the Peruvian port city of El Callao. She shows

how ... mineralized bodies assemble in conjunction with the port's infrastructure of transport and storage ... : depositories, trucks, and unpaved roads that move metals from the Andes to the sea but that also collect, leak, and disperse rogue particulates that passing bodies assimilate.

(p. 23)

Human bodies act as storage vessels for the lead seepage that would otherwise be too costly to contain or remediate. By seeing bodies as storage infrastructure, Graeter shows us the stakes of a material analysis that puts humans and non-humans within the same frame.

While these studies focus on the large and mega-scales of infrastructure, including entire national and regional assemblages of extraction, transport, and storage, Sabine Luning and Robert Pijpers (2017) focus on infrastructure at the smaller scales of artisanal mining and mining exploration. The former is often thought of as being entirely chaotic in spatial organization, while the latter conjures romantic notions of lone prospector geologists roaming the hills. Yet, an infrastructural approach, particularly a three-dimensional one that goes underground and not only across the surface, helps us understand some of the tense issues surrounding access to ore. In the context of gold mining in Ghana, the authors show how access is "informed by an intricate combination of social and natural factors: a socio-political field of relations anchored onto an underground of geological structures" (p. 762). Working with artisanal miners in Northern Ghana, Esther Van De Camp also employs a three-dimensional perspective, with particular attention to how gold's properties as a substance condition the infrastructures and governance patterns above and below ground (2016). These studies demonstrate compellingly how the material is always also political, and vice versa.

Waste

Along with the materially oriented work on infrastructure, new scholarship focuses on the substances that are produced or emerge as waste or by-products of extraction. This attention reflects a willingness to provide descriptions and analyses that run counter to or provide alternatives to the image put forward in public-facing representations of mining and extraction produced by corporations and government entities committed to extractive developmentalism (in which waste and its effects are typically erased or downplayed, when they are not turned into profit).

In her book on the uranium economy in the Navajo Nation, Traci Brynne Voyles brings into focus the process of what she calls "wastelanding" (2015). Wastelanding refers to patterns of words and actions that, as Voyles puts it, "render certain bodies and landscapes pollutable" (p. 51). Voyles' work resonates with Graeter's, described above; as with lead, uranium extraction is a particularly apt example of this isomorphic relationship between land and bodies, because of the multiscalar characteristics of radioactive materials, which affect the atmosphere, earth, air, and human body all the way down to the molecular level.

Voyles' book counterposes waste and value, but not all anthropological work on extraction follows this dichotomy. Pablo Jaramillo examines the ways in which small-scale gold miners in Marmato, Colombia treat various kinds of "leftovers" in mining as a form of material and affective "future-making". Jaramillo writes that "no 'waste' exists in miners' eyes, only leftovers, geological objects with potential agency" (2020: 50). In an article on emerald mining, also in Colombia, Vladimir Caraballo Acuña (2019) also notes how miners extract value from *tambre,* the slurry produced in mining that still contains enough emeralds to have provided a crucial margin of livelihood for many *guacheros* (artisanal emerald miners). Similarly, Sebastián Ureta (2016) draws on Science and Technology Studies to critique technocratic and anti-political approaches to waste. Focusing on the tunnels for mine tailings from a Chilean copper mine, Ureta mobilizes a notion of "caring for waste" that accounts for human and non-human actors (including the dogs that are rescued from the tailings tunnel) and for the materials themselves.

Bodies and substance

A number of works address the sensuous and affective connections between human bodies and substances and materials in extraction. Jessica Smith Rolston (2013) describes the ways in which coal miners in Wyoming mediate between the materials of the mine and the machines used in coal extraction, noting that "labor relations take on material form as workers navigate the harms presented by equipment operation and the form, height, and pitch of highwalls" (p. 583) and drawing attention to miners' appreciation of the agency of rock and other materials. González Colonia (2012) and Jaramillo (2020) note a similar capacity to recognize the agency of mined substances in Marmato, Colombia, in that case focused particularly on the agency and "living" (*vivo*) quality of gold, often apprehended through its appearance to humans or through bodily engagement with gold in the wooden pan called a *batea*. The *batea* as a technology of bodily engagement with gold is both necessary to and iconic of small-scale gold mining (Ferry and Ferry 2017). Additionally, in southwestern and western Burkina Faso, "[a] commonly held view is that gold is a living creature that moves and can suddenly appear or disappear to the accompaniment of strange lights or sounds (Werthmann 2003: 110). In this case, gold's agency directly affects the bodies of those who come into contact with it, by drinking their blood and "eating" them (p. 110).

Sabine Luning picks up on Werthmann's analysis and especially the fact that gold can be "killed" with blood or urine, but the "corresponding body part" from which these bodily substances came will also die, interpreting these in a broader cosmological context in which gold must be killed (and in its turn, kills) in order to be transmitted from the earth – which exists beyond relations of ownership to the house, a space governed by and appropriate to such relations (2012). Luning's analysis underscores a key point: that while gold is found in all regions of the world and while all humans are bodies, the embodied relations between humans and gold must be understood within their particular social, spatial, and cosmological context.

In her ethnography on Thai gem-cutters, Annabel Vallard (2020) notes the intimate connections between cutters' bodies and the stones. The emergence of the stone as a valuable material (see below), indeed, depends on the attunement of the cutter's body to it:

> To achieve this balance and symmetry, practitioners rely on the expertise and sensibility of their hands, the touch of their skin, and the sharpness of their eyes, trained through long years of practice. … Open to the eventual emergence of material variations, invisible or unperceived at first, cutters demonstrate a strong sensibility for the affordances of the matter.
>
> *(p. 128)*

Rosalind Morris (2008) focuses on another way in which miners' bodies are caught up in the material world of South African mines: through the "hellish cavern of noise" (p. 104) in which they extract diamonds or gold, a world of sonic violence that frequently causes deafness. She writes: "The miner's ear is attuned to what will destroy him, what is already destroying him in the moment that he hears, if he hears" (p. 96). His sensory engagement with the mine frequently destroys his capacity to hear.

In another consideration of mined substances and bodily damage, Agata Mazzeo explores the embodiment of asbestos as a form of consumption:

> My interlocutors often used the expression "comer o pó" [eating the dust] to refer to asbestos exposure. "We ate food mixed with fibres", they said, and in my analysis, the expression "eating the dust" effectively evoked the "embodiment" of the processes related to asbestos capitalism.
>
> *(Mazzeo 2018: 225)*

Mazzeo's analysis echoes that of Morris in seeing embodiment as damage as a material idiom for the effects of extractive capitalism on miners, and this approach also resonates with Voyles' concept of "wastelanding" described above and with the telling title of Nash's book *We Eat the Mines and the Mines Eat Us* (1993). Ryan Cecil Jobson, in an essay on W.E.B. Du Bois' narration of the rootedness of capitalism in slavery in *Black Reconstruction in America* (2017 [1935]), notes that enslaved Africans represent a massive extraction of energy from the bodies of humans who are thought of and treated as things, and that this energy transfer accompanies and makes possible the transfer of energy based in fossil fuels. His analysis builds on that of Mitchell and others to emphasize the relationship between embodied racialized capital and fossil capital.

Values and valuation

A number of works engage questions of value and valuation in relation to mined objects and substances. For instance, Ferry elaborates a process by which the inalienable wealth of the Santa Fe Cooperative in Guanajuato, Mexico is rendered valuable for multiple collectivities through an idiom of patrimony as persisting through interchangeable substances. In this process, the subsoil patrimony of the nation (silver) shares substance with the houses, bodies of miners, and city (which are the patrimony of the family and Cooperative and after 1988, the world, through UNESCO's World Heritage City programme). The links between these corporate groups help to resolve the apparent paradox of mining silver in the Cooperative – a non-renewable, global article of trade which must also be passed down to future generations (an "inalienable commodity" Ferry 2005). Silver departs from Guanajuato but also remains in the form of miners' houses, city plazas, and the jobs of Cooperative members. Silver therefore continues to participate in projects of making, retaining, and attracting value (through raising and educating children, mining silver, or preserving the architectural patrimony of the city). D'Angelo, working with diamond miners in Sierra Leone reports a similar process; one of his interlocutors "indicated the brick and concrete building where we were, a building surrounded by many other homes in the village that were instead made of mud and wood [and said] 'These are diamonds'" (2014: 277). At the same time, for these miners, diamonds are pure exchange-value. More precisely, their use-value for consumers is hidden and its mysterious character helps to establish their exchange-value – thus reconfiguring Marx's conceptualization of the mystification of exchange-value, as here is it the use-value that is mystified. In addition to extending Walsh's work, D'Angelo's argument engages with Calvão's idea of opacity and transparency as qualia for diamonds (2020), and with Smith's consideration of the fast temporalities of coltan (2011).

Value-making occurs at many other points in extraction, such as exploration and forecasting of rubies in Greenland and copper in Ecuador (Brichet 2020; Kneas 2018). Anna Tsing's piece "Inside the Economy of Appearances" uses the limit case of the Bre-X hoax and accompanying scandal, in which thousands were bamboozled with the promise of gold in Busang, Indonesia that turned out to be non-existent, to explore the nature of financial investment as a performance:

> Journalists compared Busang, with its lines of false drilling samples, to a Hollywood set. But it was not just Busang; it was the whole investment process.

No one would ever have invested in Bre-X if it had not created a performance, a dramatic exposition of the possibilities of gold.

(Tsing 2000: 118)

Her argument underscores the ways in which the performance of value (or future value) creates value in the present (Gilbert 2020).

A number of works have looked at how gemstones are valued, particularly in stages of gem-cutting, mineral "preparation" (Ferry 2013), or consumption. These types of extractive material often seem to have a more "subjective" set of criteria for valuation than materials like oil, gold, silver, and others. Often there is a tension between systems of certification of different sorts, such as those provided by the Gemological Institute of America (Falls 2014), and ideologies of connoisseurship or affect. For instance, in his work on sapphire extraction in Madagascar Walsh notes that "every sapphire is unique and might be evaluated as such … and the price of even a single stone will never be the same twice" (2004: 228; see also Brazeal 2017; Vallard 2020). In the case of sapphires, this leads to a complex interplay between knowledge and deception, one that is rendered even more elusive by the presence in the market of lab-grown "synthetic" sapphires whose material form cannot be distinguished from "natural" sapphires produced by actual geological conditions (Walsh 2010).

Susan Falls (2014) and Filipe Calvão (2015) both explore how the value of diamonds cannot be completely contained within the system of standardization that purports to assign such value. As Calvão describes it, "the materiality of a diamond – its thingness – remains at the centrepiece of negotiations over value," while at the same time, "diamonds also emerge as an unstable material by way of a ritualized negotiation" (p. 193). That is, the material, from the moment of its emergence as such – as a diamond, for example – is already entangled with processes of valuation that helped to generate it, and the shifting fields of valuation in gemstone extraction, cutting, and trade show this especially well.

4 Future directions

The centrality of materials and substances in anthropology continues to gain momentum and to extend itself into new domains and topics of inquiry, including those related to extraction. Emerging work on those areas that are sometimes experientially opposed to the world of the material, by interlocutors and also by scholars, such as financial and digital worlds (Arboleda 2020; Gilbert 2020), opens new pathways for scholarship. Scholars of these areas have begun to attend to their material dimensions (Starosielski 2015), and newer work is focusing more on the extracted materials and substances that make up these material assemblages (see Smith and Mantz 2014 for an earlier example). Examples would include work on gold as a physical substance that is also a financial technology (Ferry 2020), the metals that make up the servers used in cloud computing, or the blockchain (Calvão and Gronwald 2019), and potentially drone and VR/AR technologies used in mining and in securing mineral commodity chains.

In addition, anthropologists are extending our studies of extractive industries into new spaces, following the extension of the industries themselves. We have begun to look at the underground; new realms of extraction are opening undersea (Pulu 2013) and in space and these projects and plans have their own distinctive materialities and employ extracted substances in their own particular ways.

The test of a theoretical "turn" and of any social theory, is their ability to provide accounts that show aspects of the world invisible or obscure in other accounts. By this measure the activation of substances and materiality as dimensions of resource extraction is a smash hit. It has allowed anthropologists to include far richer descriptions of the process of extraction, the lives, bodies, sense of miners, cutters, traders, and consumers, and to provide highly productive theorizations of the enmeshment of the social and material. Prices are up and down; it is to be hoped that anthropologists' attention to the material pasts, presents, and futures of extraction will persist.

Notes

1 In contrast to some scholars (cf. Ingold 2012).
2 While a complete discussion of such approaches lies beyond what a short chapter like this can do, they include those that see things as "material culture" (Geismar 2013; Miller 1987), those that see things as imposing their own impingements on humans and the world, so that form takes shape through the forces of materials (Barad 2003; Bennett 2010; Deleuze and Guattari 1988; Ingold 2012), and a range of semiotic approaches (Chumley and Harkness 2013; Law 2009; Munn 1992).
3 In 2018, Jacka writes that "of deep consternation to the people I work with in the eastern Porgera Valley [near Mt. Kare] is what would happen to the 'diamond python' – the source of all minerals and petroleum in the PNG highlands ... now that foreign, corporate powers held the true source of mineral wealth" (2018: 68).
4 Rogers gives an excellent summary of literature on the qualities and material and discursive affordances of oil (2012: 287–288).
5 While Ortiz's classic work *Cuban Counterpoint: Tobacco and Sugar* (1995) is not usually explicitly invoked, it stands as an early exemplar of a relational approach to the study of materials.

Bibliography

Abercrombie T.A. (2016) The iterated mountain: Things as signs in Potosí. *The Journal of Latin American and Caribbean Anthropology* 21 (1): 83–108.

Adunbi O. (2013) *Oil wealth and insurgency in Nigeria.* Bloomington: Indiana University Press.

Akong C. (2020) Reframing matter: Towards a material-discursive framework for Africa's minerals. *The Extractive Industries and Society* 7 (2): 461–469.

Appel H. (2012) Offshore work: Oil, modularity, and the how of capitalism in Equatorial Guinea. *American Ethnologist* 39 (4): 692–709.

Appel H., Mason A. and Watts M. (2015) *Subterranean estates: Life worlds of oil and gas.* Ithaca: Cornell University Press.

Apter A. (2008) *The pan-African nation: Oil and the spectacle of culture in Nigeria.* Chicago: University of Chicago Press.

Arboleda M. (2020) *Planetary mine. Territories of extraction under late capitalism.* London: Verso Press.

Ballard C. and Banks G. (2003) Resource wars: The anthropology of mining. *Annual Review of Anthropology* 32 (1): 287–313.

Barad K. (2003) Posthumanist performativity: Toward an understanding of how matter comes to matter. *Signs: Journal of Women in Culture and Society* 28 (3): 801–831.

Barry A. (2013) *Material politics: Disputes along the pipeline*. London: John Wiley & Sons.

Bennett, J. (2010) *Vibrant matter: A political ecology of things*. Durham: Duke University Press.

Biersack A. (1999) The Mount Kare python and his gold: Totemism and ecology in the Papua New Guinea highlands. *American Anthropologist* 101 (1): 68–87.

Brazeal B. (2017) Austerity, luxury and uncertainty in the Indian emerald trade. *Journal of Material Culture* 22 (4): 437–452.

Brichet N. (2020) A piece of Greenland? Making marketable and artisan gemstones. *Anthropological Journal of European Cultures* 29 (1): 80–100.

Calvão F. (2020) Transparent minerals and opaque diamond sources. In: Ferry E., Vallard A. and Walsh A. (eds) *The anthropology of precious minerals*. Toronto: University of Toronto Press, 147–163.

Calvão F. (2013) The transporter, the agitator, and the kamanguista: Qualia and the in/visible materiality of diamonds. *Anthropological Theory* 13 (1–2): 119–136.

Calvão F. (2015) Diamonds, machines, and colours: Moving materials in ritual exchange. In: Drazin A. and Küchler S. (eds) *The social life of materials: Studies in materials and society*, London: Bloomsbury Publishing, 193–208.

Calvão F. and Gronwald V. (2019) *Blockchain in the mining industry: Implications for sustainable development in Africa*. South African Institute of International Affairs: African Perspectives, Global Insights.

Caraballo AcuñaV. (2019) Semiotic distribution of responsibility: An ethnography of overburden in Colombia's emerald economy. *The Extractive Industries and Society* 6 (4): 1040–1046.

Chowdhury N.S. (2016) Mines and signs: Resource and political futures in Bangladesh. *Journal of the Royal Anthropological Institute* 22 (S1): 87–107.

Chumley L.H. and Harkness N. (2013) Introduction: Qualia. *Anthropological Theory* 13 (1–2):3–11.

Coronil, F. (1997) *The magical state: Nature, money, and modernity in Venezuela*. Chicago: University of Chicago Press.

Davidov V. (2014) Land, copper, flora: Dominant materialities and the making of Ecuadorian resource environments. *Anthropological Quarterly* 87 (1): 31–58.

D'Angelo L. (2014) Who owns the diamonds? The occult eco-nomy of diamond mining in Sierra Leone. *Africa* 84 (2): 269–293.

Deleuze G. and Guattari F. (1988) *A thousand plateaus: Capitalism and schizophrenia*. London: Bloomsbury Publishing.

Du Bois W.E.B (1998 [1935]) *Black reconstruction in America, 1860–1880*. New York: Free Press.

Epstein A.L. (1958) *Politics in an urban African community*. Manchester: Manchester University Press.

Falls S. (2014) *Clarity, cut, and culture: The many meanings of diamonds*. New York: NYU Press.

Ferguson J. (2005) Seeing like an oil company: Space, security, and global capital in neo-liberal Africa. *American Anthropologist* 107 (3): 377–382.

Ferry E.E. (2005) *Not ours alone: Patrimony, value, and collectivity in contemporary Mexico*. New York: Columbia University Press.

Ferry E.E. (2013) *Minerals, collecting, and value across the US–Mexico border*. Bloomington: Indiana University Press.

Ferry E. (2016) On not being a sign: Gold's semiotic claims. *Signs and Society* 4 (1): 57–79.

Ferry E. (2020) Speculative substance: "Physical gold" in finance. *Economy and Society* 49 (1): 92–115.

Ferry E. and Ferry S. (2017) *La Batea*. Bogotá: Editorial Icono; New York: Red Hook Editions.

Ferry E.E. and Limbert M.E. (2008) *Timely assets: The politics of resources and their temporalities*. Santa Fe: School of Advanced Research Press.

Field L. (2012) The gold system: Explorations of the (ongoing) fate of Colombia's pre-Columbian gold artifacts. *Antipoda. Revista de Antropología y Arqueología* 14: 67–94.

Geismar H. (2013) *Treasured possessions: Indigenous interventions into cultural and intellectual property*. Durham: Duke University Press.

Gibson J.J. (2014) *The ecological approach to visual perception*. New York: Psychology Press.

Gilbert R. (2020) Speculating on sovereignty: "Money mining" and corporate foreign policy at the extractive industry frontier. *Economy and Society* 49 (1): 16–44.

González Colonia C.J. (2012) *Brujería, minería tradicional y capitalismo transnacional en los Andes colombianos. El caso del pueblo minero de Marmato-Colombia*. Buenos Aires: Universidad Nacional de San Martin.

Gluckman M. (1944) Anthropological problems arising from the African industrial revolution. In: *Social change in modern Africa*. Oxford: Oxford University Press for the International African Institute.

Graeter S. (2020) Infrastructural incorporations: Toxic storage, corporate indemnity, and ethical deferral in Peru's neoextractive era. *American Anthropologist* 122 (1): 21–36.

Harris O. (1989) The earth and the state: The sources and meanings of money in Northern Potosi, Bolivia. In: Bloch M. and Parry J. (eds) *Money and the morality of exchange*, Cambridge: Cambridge University Press, 232–268.

Hitchcock P. (2015). Velocity and viscosity. In: Appel H., Mason A. and Watts M. (eds) *Subterranean estates: Life worlds of oil and gas*, Ithaca: Cornell University Press, 45–60.

Ingold T. (2012) Toward an ecology of materials. *Annual Review of Anthropology* 41: 427–442.

Ingold T. (2018) Back to the future with the theory of affordances. *HAU: Journal of Ethnographic Theory* 8 (1–2): 39–44.

Jacka J.K. (2018) The anthropology of mining: The social and environmental impacts of resource extraction in the mineral age. *Annual Review of Anthropology* 47: 61–77.

Jaramillo P. (2020) Mining leftovers: Making futures on the margins of capitalism. *Cultural Anthropology* 35 (1): 48–73.

Jobson, R.C. (2021) Dead labor: On racial capital and fossil capital. In: Jenkins D. and Leroy J. (eds) *Histories of racial capitalism*, New York: Columbia University Press, 215–230.

Jorgensen D. (1998) Whose nature? Invading bush spirits, travelling ancestors, and mining in Telefolmin. *Social Analysis: The International Journal of Social and Cultural Practice* 42 (3): 100–116.

Keane W. (2018) Perspectives on affordances, or the anthropologically real: The 2018 Daryll Forde Lecture. *HAU: Journal of Ethnographic Theory* 8 (1–2):27–38.

Kirsch S. (2002) Anthropology and advocacy: A case study of the campaign against the Ok Tedi mine. *Critique of Anthropology* 22 (2): 175–200.

Kneas D. (2018) From dearth to El Dorado: Andean nature, plate tectonics, and the ontologies of Ecuadorian resource wealth. *Engaging Science, Technology, and Society* 4: 131–154.

Larkin B. (2013) The politics and poetics of infrastructure. *Annual Review of Anthropology* 42: 327–343.

Latour B. (2012) *We have never been modern*, Cambridge, MA: Harvard University Press.

Law J.L. (2009) Actor network theory and material semiotics. In: Turner B.S. (ed.) *The new Blackwell companion to social theory.* Hoboken, NJ: Blackwell Publishing, 141–158.

Li F. (2015) *Unearthing conflict: Corporate mining, activism and expertise in Peru.* Durham: Duke University Press.

Limbert M. (2010) *In the time of oil: Piety, memory, and social life in an Omani town,* Palo Alto: Stanford University Press.

Limbert M. (2016) Liquid Oman: Oil, water, and causality in Southern Arabia. *Journal of the Royal Anthropological Institute* 22 (S1): 147–162.

Luning S. (2012) Gold, cosmology, and change in Burkina Faso. In: Panella C. (ed.) *Lives in motion, indeed. Interdisciplinary perspectives on social change in honour of Danielle de Lame.* Tervuren: Royal Museum for Central Africa, 323–340.

Luning, S. and Pijpers, R.J. (2017) Governing access to gold in Ghana: In-depth geopolitics on mining concessions. *Africa* 87 (4): 758–779.

Mantz J.W. (2008) Improvisational economies: Coltan production in the eastern Congo. *Social Anthropology* 16 (1): 34–50.

Maurer B., Nelms T.C. and Swartz L. (2013) "When perhaps the real problem is money itself!": The practical materiality of Bitcoin. *Social Semiotics* 23 (2): 261–277.

Mazzeo A. 2018) The temporalities of asbestos mining and community activism. *The Extractive Industries and Society* 5 (2): 223–229.

Miller D. (1987) *Material culture and mass consumption,* London: Blackwell Press.

Mitchell T. (2011) *Carbon democracy: Political power in the age of oil,* London: Verso Books.

Mol A. (2002) *The body multiple: Ontology in medical practice,* Durham: Duke University Press.

Moodie T.D. and Ndatshe V. (1994) *Going for gold: Men, mines, and migration,* Berkeley: University of California Press.

Moretti D. (2006) The gender of the gold: An ethnographic and historical account of women's involvement in artisanal and small-scale mining in Mount Kaindi, Papua New Guinea. *Oceania* 76 (2): 133–149.

Morris R.C. (2008) The miner's ear. *Transition* 98: 96–115.

Munn N.D. (1992) *The fame of Gawa: A symbolic study of value transformation in a Massim (Papua New Guinea) society,* Durham: Duke University Press.

Nash J.C. (1993) *We eat the mines and the mines eat us: Dependency and exploitation in Bolivian tin mines,* New York: Columbia University Press.

Ortiz F. (1995 [1940]) *Cuban counterpoint, tobacco and sugar,* Durham: Duke University Press.

Parmentier R.J. (1994) *Signs in society: Studies in semiotic anthropology,* Bloomington: Indiana University Press.

Pijpers R.J. (2011) When diamonds go bust: Contextualising livelihood changes in rural Sierra Leone. *Journal of International Development* 23 (8): 1068–1079.

Powdermaker H. (1962) *Copper town: Changing Africa: The human situation on the Rhodesian Copperbelt,* New York: Harper & Row.

Pulu T.B. (2013) Deep sea tension: The Kingdom of Tonga and deep sea minerals. *Te Kaharoa* 6 (1): 50–76.

Richardson T. and Weszkalnys G. (2014) Introduction: Resource materialities. *Anthropological Quarterly* 87 (1): 5–30.

Rogers D. (2012) The materiality of the corporation: Oil, gas, and corporate social technologies in the remaking of a Russian region. *American Ethnologist* 39 (2): 284–296.

Rogers D. (2015) Oil and anthropology. *Annual Review of Anthropology* 44: 365–380.

Rolston J.S. (2013) The politics of pits and the materiality of mine labor: Making natural resources in the American West. *American Anthropologist* 115 (4): 582–594.

Sawyer S. (2004) *Crude chronicles: Indigenous politics, multinational oil, and neoliberalism in Ecuador,* Durham: Duke University Press.

Sheller M. (2014) *Aluminum dreams: The making of light modernity*, Cambridge, MA: MIT Press.

Smith J.H. (2011) Tantalus in the digital age: Coltan ore, temporal dispossession, and "movement" in the Eastern Democratic Republic of the Congo. *American Ethnologist*, 38 (1): 17–35.

Smith J.H. and Mantz J.W. (2014). Do cellular phones dream of civil war? The mystification of production and the consequences of technology fetishism in the Eastern Congo. In: Kirsch M. (ed.) *Inclusion and exclusion in the global arena*. New York: Routledge, 83–106.

Starosielski N. (2015) *The undersea network*, Durham: Duke University Press.

Taussig M. (2009) *My cocaine museum*, Chicago: University of Chicago Press.

Tsing A.L. (2000) Inside the economy of appearances. *Public Culture* 12 (1): 115–144.

Ureta S. (2016) Caring for waste: Handling tailings in a Chilean copper mine. *Environment and Planning A* 48 (8): 1532–1548.

Vallard A (2020). When stones become gems. Valuations of minerals in Thailand. In: Ferry E., Vallard A. and Walsh A. (eds) *The anthropology of precious minerals*. Toronto: University of Toronto Press, 118–146.

Van de Camp E. (2016) Artisanal gold mining in Kejetia (Tongo, northern Ghana): A three-dimensional perspective. *Third World Thematics: A TWQ Journal* 1 (2): 267–283.

Voyles T.B. (2015) *Wastelanding: Legacies of uranium mining in Navajo country*, Minneapolis: University of Minnesota Press.

Walsh A. (2004) In the wake of things: Speculating in and about sapphires in Northern Madagascar. *American Anthropologist* 106 (2): 225–237.

Walsh A. (2010) The commodification of fetishes: Telling the difference between natural and synthetic sapphires. *American Ethnologist* 37 (1): 98–114.

Watts M.J. (2012) A tale of two gulfs: Life, death, and dispossession along two oil frontiers. *American Quarterly* 64 (3): 437–467.

Werthmann K. (2003) Cowries, gold and "bitter money": Gold-mining and notions of ill-gotten wealth in Burkina Faso. *Paideuma* 49: 105–124.

Wilson G. (1941–1942) *An essay on the economics of detribalization in Northern Rhodesia, Parts I and II*. Rhodes-Livingstone Papers 5 and 6. Manchester: Manchester University Press for the Institute for Social Research, University of Zambia.

7

MINEWORKERS

Benjamin Rubbers

1 Introduction

To use Marx's famous concept, the mine could be regarded as a "hidden abode" of the modern global economy. In this highly securitized space, inside which "there is no admittance except on business", minerals are extracted from the earth's crust and transformed into commodities such as copper, coal, or gold (Marx 1887: 123). While this process has historically involved some of the worst forms of labour exploitation, most end-consumers know very little about it, as though these commodities have an intrinsic value entirely determined by the laws of supply and demand in international markets.

A number of anthropologists have shed light on this hidden abode of production – the labour process through which mineral commodities are produced – using an ethnographic lens.[1] This chapter reviews this literature, taking as its point of departure a selection of monographs that have been influential in the study of industrial mineworkers since the 1950s. The first section considers Arnold Epstein's and Hortense Powdermaker's pioneering studies on Northern Rhodesia (now Zambia) in the 1950s and the reappraisal of their work by James Ferguson and Hugh Macmillan in the 1990s. The second section discusses the works of Robert Gordon on Namibia, of June Nash on Bolivia, and of Jane Finn on Chile and the United States, in the last quarter of the twentieth century. Although it also deals with other recent studies, the final section showcases Dinah Rajak's and Jessica Rolston's research on South Africa and the United States in the 2000s, respectively. These monographs can be read as indicating major shifts in the questions, themes, and approaches of the anthropology of mineworkers since the 1950s.

As the questions and themes addressed by these landmark studies are related to the organization of the mining industry in a given region at a certain period of history, each section below starts with a brief presentation of their context, before

DOI: 10.4324/9781003018018-7

turning to the insights they offer on labour in the mining industry. To my knowledge, a consideration of such studies in a comparative and cumulative perspective, as a research tradition in its own right, is not common practice in mining studies. If some of these studies are often cited in this literature, they are rarely compared to each other, and most authors in the field aim primarily to contribute to regional research and/or to broader theoretical debates in their discipline, not to the study of labour in mining as such. Such an endeavour nevertheless allows the growth of a better understanding of the different approaches that have been developed to study mineworkers and, more generally, to highlight the benefits that can be drawn from taking a labour perspective in the field of mining studies.

2 Adaptation to modernity

In the 1930s, new underground mines were developed in the Copperbelt of Northern Rhodesia, now the Republic of Zambia. Following the model of the South African mining industry, the companies operating these mines initially recruited single migrant men for short work contracts. But the increase of production, the shortage of experienced labour, and the first strikes by African workers pushed them to gradually encourage workforce stabilization and to build company towns capable of accommodating workers with their wives and children. This process was already well underway when Epstein and Powdermaker conducted their research in Luanshya in 1953–1954. The monographs that emerged from their respective research reflect the new interest of social scientists in the consequences of industrialization and urbanization in the colonies of the time, and the problem of migrants' adaptation to the "modern" world of the city. With thousands of workers from rural areas migrating for increasingly longer periods of time to the towns that had been built around the mines, the Copperbelt was seen as a fantastic "laboratory" in which to study social change (Schumaker 2001).

As a member of the Rhodes-Livingstone Institute (RLI), Arnold Epstein was influenced by Max Gluckman, who had been the director of this research centre from 1942 to 1947. Building on the structural functionalism of Alfred Radcliffe-Brown and Edward Evans-Pritchard, Gluckman (1945) looked at the colonial city not as a cultural space foreign to Africans, but as a social situation to which they adapted their behaviour. Although this approach did not challenge the colonial assumption that Africans belonged to "tribes" – that is, to well-defined social units – in rural areas, it suggested that what this meant changed with their migration to town.

While it does not focus on mineworkers exclusively, Epstein's *Politics in an Urban African Community* (1958) gives prominent place to the processes that led to the emergence of the African Mineworkers' Union in 1949. At the beginning of the mining industry in the 1930s, colonial companies had established a system based on "tribal elders" to manage their relationships with workers, who came from different tribes all over Central Africa. The elders were responsible for conveying the compound manager's orders down to the workers from their tribe, and for taking these workers' requests up to the compound manager. They also provided moral guidance to tribe members and settled any conflicts that arose among them in the compound. The authority of these elders, who were elected by the members of their tribe, was based, above all, on their age, wisdom, and proximity to the chief at home.

This "traditional" form of authority, Epstein explains, was put into question by strikes in 1935 and 1940 during which the elders played into the hands of management and became the focus of a growing suspicion among the workers. They were progressively replaced by a new generation of young urban leaders who held well-paying jobs, spoke English fluently, and had a lifestyle closer to that of the Europeans. From the workers' point of view, such characteristics gave them more legitimacy to negotiate the terms of employment and living conditions with employers. In the aftermath of the Second World War, these new leaders successfully participated in the creation of an African trade union and campaigned to abolish the system of tribal elders.

In Epstein's analysis, however, the replacement of tribal elders with urban leaders did not dissipate tribal allegiance in mine townships. If the workers put their tribal identities in brackets when confronting the Europeans, these continued to influence everyday social life, and provided a political base in situations of competition between Africans. Tribalism – which anthropologists would later call ethnicity – thus remained a key factor in union elections. Since it was based on broader categories than those prevailing in rural areas, and used only in certain circumstances, this urban tribalism was not to be conceived as a remnant of the traditional past. Far from being a simple reflection of rural tribalism, Epstein argued, it was to be interpreted in light of the class differentiation that was taking form within the workforce, and of the complexity of urban life.[2]

In contrast to Epstein, Hortense Powdermaker was not a member of the RLI, and took a different approach to the changes experienced by African mineworkers in the Copperbelt. Although she studied anthropology in Great Britain, once back in the United States she distanced herself from the structural functionalism of Radcliffe-Brown to develop an approach more influenced by the tradition of American cultural anthropology (Hier and Kemp 2002). In her view, the problem with structural functionalism was not so much that it was unable to account for social change as that it ignored cultural dynamics and the psychological causes of human behaviour. Another difference with Epstein is that, when Powdermaker arrived in Northern Rhodesia in 1953, she obtained the authorization to conduct her research in the mine township. As a result, her book *Copper Town: Changing Africa* (1962) is more ethnographic: while it is limited to the township, it describes the everyday life of its inhabitants in more detail than Epstein's work. Generally, through her description of Black/White interactions and workers' leisure activities, Powdermaker provides insights into the experiences of African workers in the Copperbelt of the 1950s that Epstein and other RLI researchers had neglected at the time (Hansen 1991).

Regarding trade unionism, if *Copper Town* broadly supports Epstein's interpretation (1958), it allows for a better understanding of the union's place in workers' lives and highlights workers' ambiguous relationship with management. While Epstein positions management and the union, Europeans and Africans, as opponents, Powdermaker (1962: chap. 7) suggests that workers appreciated different aspects of the company's paternalism and that, in various circumstances such as football games and English lessons, the etiquette presiding over interactions between Africans and Europeans was made less important. For her, although management and the union were on opposing sides during strikes, in daily life they were comparable symbols of benevolent authority (p. 106).

From a broader perspective, Powdermaker, like Epstein, looks at the social changes taking place in the Copperbelt in the 1950s as a process analogous to the Industrial Revolution in nineteenth-century Europe. Where she diverges from Epstein is in her analysis of this process in terms of psychological adaptation to a new modern culture, rather than the tensions characteristic of the colonial social structure. This is one of the reasons why *Copper Town* is cited in the literature less often than *Politics in an Urban African Community*, and why it does not account with the same degree of detail for the rise of new social and political aspirations among the urban Africans who would eventually play a significant role in the struggle for the independence of Zambia in 1964 (Cooper 1996; Larmer 2007). Moreover, Powdermaker's work cannot in any simple way be included in RLI's body of research, which would later be represented as a fully-fledged "school" (the Manchester school) in the history of British social anthropology (Evens and Handelman 2006; Werbner 1984).

Together with other studies from the RLI, the monographs of Epstein and Powdermaker contributed to place a series of themes – ethnicity, trade unions, marriage, leisure, and aspirations to modernity – at the heart of the anthropology of mineworkers. As we will see, these themes were taken up in subsequent studies on the subject. What distinguishes the work of these early scholars is that in addressing these themes, they took an approach that highlighted the reinvention of traditions in urban settings, and the capacity of African workers to adapt to colonial modernity.

In the 1990s, Ferguson (1990a, 1990b, 1999) criticized the modernist narrative underlying this body of research. His critique was based on interviews with mineworkers in 1985–1986, in a context marked by the dramatic decline of the Zambian mining industry. Such a decline challenged the "expectations of modernity" that social scientists and the workers themselves had in the past. While researchers since the 1950s predicted that the African urban population would grow and become more settled, most of the mineworkers Ferguson interviewed in the 1980s actually planned to retire in rural areas. Zambia's economic crisis was thus to be understood at the same time as a crisis in how modernity was represented and understood.

The interest of Ferguson's research lies less in the (relatively scant) ethnographic data his publications provide about mineworkers in the 1980s than in his reappraisal of the dualistic and teleological approach to rural/urban migration that anthropologists developed in the 1950s. In his critique, this approach was not only contradicted by the mining industry's subsequent decline, but also led the anthropologists to overlook the diversity of practices and discourses at the time of their fieldwork in the late colonial period. A good illustration is provided by Powdermaker's and Epstein's analyses of the evolution of the family in the 1950s (Powdermaker 1962; Epstein 1981; see Ferguson 1999: 166–206). Both acknowledged that African households, which were characterized by strong suspicion between spouses, high rates of divorce, and the presence of distant relatives in the home, were far from the model of the modern nuclear family promoted by colonial authorities. Both authors were nevertheless convinced that this modern nuclear family was rapidly gaining ground with the stabilization of the African population in towns. Yet this type of family was still far from being common when Ferguson completed his fieldwork. In the 1980s, family life among mineworkers maintained very much the same characteristics as those Powdermaker and Epstein had described in the 1950s.

Later, Ferguson's reappraisal of early studies on the Copperbelt was itself criticized by historian Hugh Macmillan (1993, 1996). He argued that these studies had been dominated by a "cyclical" view of history rather than a "progressive" one, and that their insistence on the increasing tendency towards urban permanence was indeed correct. Moreover, he contended that past research had never denied the continuing importance of rural–urban links. If this critique is biased in several respects (see Ferguson 1994), it nevertheless has the merit of drawing further attention to the temporalities of the mining industry, its booms and busts. Such variations confront anthropologists with the risk of lapsing into projections that are at times optimistic (in terms of development and inclusion) and at other times pessimistic (in terms of decline and abjection). To overcome this problem, it is essential that researchers keep the fragile nature of mining investment in mind and focus on how people themselves imagine and prepare for their future (or not) in connection with their experience of the past (see D'Angelo and Pijpers 2018; Ferry and Limbert 2008). In a sense, the hopes and fears that the future raises underlie all the social relationships implicated by mining activities.

3 Cultures of resistance

Unlike the studies reviewed in the first section, the main focus of the three monographs of interest to us here – *Mines, Masters and Migrants* by Gordon (1977), *We Eat the Mines and the Mines Eat Us* by Nash (1993 [1979]) and *Tracing the Veins* by Finn (1998) – is not on mineworkers' aspirations to modernity, but on their capacity to create their own worlds in the mines. As we will see, this perspective leads these authors to put more emphasis on resistance, solidarity, and dependency than Epstein or Powdermaker.

In the late nineteenth century, South African mines implemented a migrant labour system that remained in place until the end of Apartheid. Although this policy underwent modifications, mining companies never adopted a stabilization policy comparable to that of Northern Rhodesia. Until the 1970s, the bulk of the workforce comprised unskilled and low-paid single men. During their stays in the mines, which rarely exceeded two years, they were housed in compounds equipped with dormitories, canteens, and basic leisure facilities. In these compounds, as in the workplace, they were subject to the discretion of white supervisors and various categories of black assistants, who did not hesitate to rely on assault and repression as management strategies.

Based on his experience as a personnel officer in 1973–1974 in a small mine near Windhoek, Namibia, Robert Gordon's (1977) book not only provided new insights into the world of black miners under Apartheid but also brought to light aspects that had been neglected in the literature on Zambia: how miners resisted the control exerted by white supervisors, and instrumentalized their relationship with them – what James Scott (1990) would call their hidden transcripts. Although the Namibian mineworkers were not formally organized, they contested management's ambition to exert total control over the compound through various forms of everyday resistance including desertion, sabotage, slowdowns, and work stoppages. To do so, they relied on the strong solidarity between black "brothers", which was fostered on a day-to-day basis through the sharing of alcohol and the exchange of gifts and services.

Drawing inspiration from Erving Goffman (1990 [1959]), Gordon showed how black mineworkers played with the rules of etiquette presiding over interactions with Whites to stay out of trouble and obtain favours. Such interactions, he explained, must be understood in the light of the reciprocal relationship that existed between Whites and Blacks in the mine's social order. On the one hand, Whites were aware that they benefited from the work of the black workers under their supervision. On the other, black workers sought the patronage of powerful Whites to access specific functions or to organize the informal trafficking of prohibited items in the compound. According to Gordon, their main purpose was to earn money to support their rural families. In contrast to Zambian workers, Namibian black workers' migration to the mines was by no means a choice that could lead them to establish themselves permanently in the vicinity of the mine. Their movement was primarily to be understood from the adverse living conditions they endured in their rural homelands under the Apartheid regime.

The Bolivian tin mines where June Nash did her fieldwork in 1969–1970 introduce a very different political context. First developed by an American company, the mines near Oruro were nationalized in 1952. While the workers continued to live together with their families in mining encampments, the nationalization of the mines was accompanied by a continuous decline in real wages. This decline in living conditions, together with the replacement of expatriates with national supervisors, pushed the mineworkers to organize recurrent strike actions despite government repression. Although the workers of this state-owned enterprise barely represented 2% of the national workforce, they were seen as the most active segment of the Bolivian labour movement. Throughout the book, Nash explores the roots of this militancy and, in doing so, develops a new approach to mineworkers' class culture.[3] Compared to the studies reviewed above, this approach innovates on at least three points.

First, Nash does not consider traditions inherited from the past as obstacles, but as resources for militant action. Particular emphasis is put on the role of pre-conquest rituals and beliefs in the making of mineworkers' class consciousness. For the author, their identity drew its strength as much from a shared popular culture – the *Cholo* culture – as from socialist ideologies.

Second, Nash shows that mineworkers' solidarity originated not only in the strong social ties they developed in the mines and the camp, but also in those from home. Wives and consumption practices were important factors in the outbreak of labour disputes. Strikes often started in the queue that women formed in front of the *pulperia* (the company store). When faced with a shortage of essential goods at home, they did not hesitate to march to demand better living conditions from management.

Finally, workers' militancy was not separated from the political economy of mining. Nash shows how the organizational, economic, and political changes that took place after the nationalization of the company in 1952 contributed to broaden the ties between underground miners, fuel their hostility towards supervisors, and significantly increase the number of labour disputes. This growing militancy was, however, brutally crushed following the 1971 Banzer military coup, one year after Nash's departure from the country.

If their culture provided Bolivian mineworkers with a "generative" base to adapt to dehumanizing working conditions and to mobilize to defend their interests, the fact remains that they depended on employment in the mines for their livelihoods. Mineworkers were fully aware of this contradiction, which they expressed in various forms, including through religion. They offered sacrifices to the spirit of the mine to protect them from accidents and allow them to continue extracting the ore. As one worker said during one of these rituals, an observation from which the book takes its title, "We eat the mines and the mines eat us." As Nash writes (1993: ix), this sentiment encapsulates "the sense miners have of their dependency on the mines in order to make a living and their sense of exploitation, both in the loss of health from lung disease and in the low returns of their labor".

How mineworkers experience this contradiction is at the heart of *Tracing the Veins* by Jane Finn (1998). This book is on the history of Butte in the United States, and Chuquicamata in Chile – two company towns that were dominated by the mining giant Anaconda from the 1920s into the 1970s. Like Nash, Finn focuses on mineworkers' agency, how they crafted the everyday, with a particular emphasis on the role of women and the solidarity that existed within the mining community. Influenced by the postmodern turn in anthropology, *Tracing the Veins* distances itself from the universal teleological narrative implicit in Nash's analysis.[4] For Finn, workers' agency did not derive its strength from a common indigenous cultural legacy that provided resources for action, but from their capacity to appropriate and divert company labour control strategies. More weight is thus given to how the corporation shaped workers' sense of class, community, and gender:[5]

As the Anaconda Company was building a steady and reliable workforce, residents of Butte and Chuquicamata were crafting their own sense of belonging. Yet neither their visions of community nor their sense of themselves as members was entirely of their own making. The corporate imagination carved the physical and social contours of community in Butte and Chuquicamata. In the process, it forged forceful images of what it meant to be good workers, citizens, and consumers.

(Finn 1998: 73)

From this perspective, the ties and traditions that united workers were not necessarily involved in the development of working-class consciousness. They resulted from lived experiences that were ambiguous and contingent, and which must be understood in light of the changing relationships between capital and labour in the course of history. The solidarity between workers sometimes enabled them to be victorious and improve their living conditions. Since they depended on the company to make a living, however, the last word belonged to capital. Following Nash, Finn thus places the contradictions of capitalism at the heart of the workers' experience. Drawing inspiration from Geertz (1973), she suggests that they were caught in a "deep play, a game one seemingly cannot win but can't not play" (Finn 1998: 219).

The developments discussed in the final chapters of the book support this pessimistic view. In the 1990s, while ex-miners in both Butte and Chuquicamata recounted how their health, as well as the places that were meaningful to them, had been gradually "consumed" by mining operations, the mining companies were putting a more flexible labour regime in place that broke in many ways with the "paternalism" of the past. As we will see in the next section, this shift foreshadowed what was going to take place in the following decades with the expansion of foreign investments in the industry and the development of new mining projects all over the world.

4 The work of mining capitalism

The period from the 1990s to the present has been marked by an increase in foreign investments in the mining sector, especially in the Global South, and the development of new mining, processing, and management techniques, which have allowed investors to mine larger deposits – including a growing share of surface mines – with fewer workers than in the past (see D'Angelo this volume). Generally speaking, the new generation of mining projects tends to be more capital-intensive than those reviewed in the two previous sections.

In this context, anthropologists' attention has been largely concentrated on the new type of relationship that mining companies have developed with local communities, with a marked interest in greenfield projects funded by companies headquartered in Western Europe or North America (e.g. Ballard and Banks 2003; Filer and Le Meur 2017; Golub 2014; Kirsch 2014; Li 2015; Rajak 2011; Welker 2014). Labour, by contrast, has tended to fade into the background as an issue of secondary importance.

This does not mean, however, that this body of literature does not provide fresh insights into the subject. In contrast to classical ethnographies, which mainly focused on workers, it sheds new light on the work of managers and experts involved in community relations management. Examining the work of company representation performed by these employees leaves behind the image of the monolithic corporation to explore in detail the government techniques that mining companies put in place, and the ways they are represented by various actors (Golub 2014; Rajak 2011; Welker 2014). Such an approach has considerable implications for the study of mining companies' labour management policies, and the relationships between management, trade unions, and workers (Rubbers 2020a).

Dinah Rajak's (2011) *In Good Company*, based on ethnographic fieldwork conducted between 2004 and 2005 on Anglo-American's Corporate Social Responsibility (CSR) policy in the platinum mines of South Africa, is particularly interesting here, as one of the targets of this policy was the company's own workforce: one of the main areas of intervention for the company within the framework of its CSR strategy in the early 2000s was to provide antiretroviral therapy to HIV-positive workers. Rajak points out the paradox of this programme, which allowed the company to present itself as a socially responsible citizen which cared for the health of its workers while limiting its responsibility in practice to a core group of permanent workers. Indeed, this programme excluded not only contract workers, who comprised a third of the workforce, but also the dependents of unskilled workers who could not afford the company health insurance. Moreover, those who were declared unfit for work for medical reasons were dismissed and handed over to the state medical services, which were unable to ensure continuity of treatment.

In Rajak's analysis, the different components of this CSR programme – nutrition, sexual life, sport, and cleanliness – share the same biopolitical purpose of improving the company's human capital and preserving it from outside sources of pollution. As such, although the company claims to break with the past, its practices can be viewed as a continuation of the corporate paternalism characteristic of the Apartheid era. The main difference is that CSR is more of a simple economic calculation. In comparison to Anglo-American corporate paternalism under Apartheid, CSR tends to be more exclusive and contingent on shareholder value.

If industrial mineworkers are no longer centre stage in the literature on mining, several anthropologists have nevertheless sought to understand how they have been affected by changes in the global mining industry since the 1990s: the rise of subcontracting, the promotion of a safety culture, the adoption of gender equality programmes, the adoption of 12-hour shifts, the development of fly-in fly-out operations, the provision of credit facilities, and/or the co-optation of trade unions as corporate partners (James and Rajak 2014; Keskula 2016, 2018; Keskula and Sanchez 2019; Rajak 2016; Rolston 2010, 2013, 2014). In contrast to ethnographies of mineworkers in the twentieth century, these studies do not seek to depict class cultures. Distancing themselves from a research agenda in terms of class-making, they focus on how workers cope with one or several of these labour management techniques.

Among these new ethnographies of mineworkers, the work of Jessica Rolston (2010, 2013, 2014) on coalminers in Wyoming, United States, is of particular interest. The daughter of a mine mechanic, Rolston worked in the mines as a summer student employee before embarking on ethnographic fieldwork in the mines around the town of Gillette from 2004 to 2012. Unlike most anthropologists, she did not only carry out her research in the mining community, or in the mining companies' offices. She also followed the workers in the mines over an entire year.

This participant observation gave her the opportunity to study what workers did and said in the workplace. Rolston emphasizes the materiality of their work, the ways in which they "sensed" the coal, the mining vehicles they drove, and the changing environment of the mine. Working in the mines led them to develop a "pit sense", a practical knowledge allowing them to navigate the hostile environment of the mines safely. Rolston's main focus, however, is the solidarity that workers developed within shift crews. Drawing inspiration from Janet Carsten (2004), she understands this solidarity as a form of workplace relatedness, which miners cultivated by working, joking, and eating together. This workplace relatedness results to a large extent from the introduction of 12-hours shifts, which led them to spend a lot of time together while living in a temporal mismatch with their families. This organization of work in crew families is key to understanding the particular form that safety and gender took in the Powder River Basin coalmines.

When these mines started operations in the late 1970s and early 1980s, the mining companies attempted to implement classical safety rules and procedures based on discipline and individual responsibility. The miners, however, minimized their participation in these safety programmes because they undermined the existing solidarity in shift crews. Instead, they set up an informal safety system through which they offered each other feedback privately without reporting to supervisors. This kinship-based safety system was eventually adopted and formalized by mining companies in the early 2000s; since then, it has allowed them to significantly improve their safety records. The miners thus succeeded in shaping mining companies' safety programmes and, by making safety a matter of collective responsibility, in creating an exception to neoliberalism.

Another peculiarity of these coalmines is that, since the beginning of their operations, their staff included a significant proportion of women who were successfully integrated into the shift crews. To understand this successful integration of women into what is generally considered a male-dominated workplace, Rolston explains that fostering a family spirit between crew members involved minimizing gender differences to promote a work ethic based on work done well, care for others, and platonic friendships with co-workers of the opposite sex. While workers did use gender stereotypes, these did not give rise to discriminatory practices against women. On the contrary, these stereotypes were used in such a way as to undo gender and to make gender difference less significant in the workplace, even though this was imperfectly realized in practice.

Finally, the workplace relatedness that developed among coalminers in the Powder River Basin helps to explain why they refused to be represented by unions. When mining companies started production in the basin, they actively sought to keep unions away from the mines by recruiting workers from agricultural communities without a union tradition and offering them higher wages and benefits than in unionized mines. Later the workers themselves voted against unionizing, fearing that the presence of unions would undermine the existing solidarity within crew families, and push management to casualize their jobs and reduce their wages and benefits. In contrast to the miners studied in the previous sections, Wyoming coalminers earned high wages and considered themselves members of the "middle class". Accordingly, they tended to be closer to management than to unions, and to defend the mining industry against its critics.

Even if mineworkers are now fewer in number and politically weaker than in the past, recent ethnographies like Rolston's show the insights that can be gained from taking work as a lens to study the transformations of the mining industry. Building on this new body of scholarship, a team of five researchers and myself have conducted ethnographic research on the "micropolitics of work" in the Congolese and Zambian copperbelts between 2016 and 2019.[6] Since the early 2000s, this region has witnessed an influx of foreign investors of various origins, who have started more than 30 new mining projects. A characteristic feature of these new mining projects is that they put in place new workforce management practices, which may include all or some of: a shift towards automation and the use of IT tools, the development of subcontracting, the entrenchment of a strict safety discipline, the professional advancement of women, the extension of credit opportunities to workers, and the organization of frequent staff downsizing operations. Our focus was on the ways new mining companies' labour practices are implemented by human resource managers (Rubbers 2020a, 2020b); the ambiguous role that trade unions play in this process (Geenen 2020; McNamara 2021a, 2021b); and how workers were affected by these changes (Musonda 2020, 2021; Pugliese 2020).

Although their distribution among mining projects is not uniform and systematic, the new flexible labour practices that foreign investors put in place collectively point to the emergence of a neoliberal labour regime which breaks in many ways from the paternalism of the state-owned enterprises that dominated the two copperbelts in the second half of the twentieth century (Rubbers 2021). However, this new labour regime was not simply imposed by foreign investors; its rise is the result of a variegated process of improvisation and adaptation involving local actors. In support of this argument, our research shows how new mining projects' labour practices were mediated, negotiated, or resisted by mineworkers, unionists, and human resource managers. It also emphasizes variations in the labour practices put in place by new mining projects depending on the type of capital involved, the type of mine being developed, and the area where they are established.

From this perspective, mining companies are not viewed as external monolithic organizations that enter into relationships with local communities (Golub 2014;

Shever 2012; Welker 2014). Instead, the mining projects they develop are studied as being themselves co-produced by various categories of actors inside and outside mining companies. From the very first steps foreign investors take in the country, they are caught in various power configurations that influence how their mining projects are developed. Accordingly, the mining boom that occurred in the Central African Copperbelt in the 2000s should not only be understood as a process of "accumulation by dispossession" (Harvey 2003; see Li 2009). It also involved a grafting process, through which mining projects became entangled in the historical trajectories of the areas where they are established in relatively new ways.

5 Conclusion

This chapter aimed to review ethnographies of industrial mineworkers as contributions to a cumulative research tradition in the anthropology of mining. Since the 1950s, the literature review presented here suggests that anthropologists have approached mineworkers from three main points of view: first, their aspirations for modernity and complex articulations of rural and urban cultures involved in the migrant labour system; second, the making of original class cultures that give mineworkers some room to resist exploitation and alienation; and, finally, the ways in which labour practices are enabled, negotiated, and resisted by various actors inside and outside companies. Thus, there has been a shift in the past two decades from a focus on workers' culture to a broader interest in the various work practices that shape the development of mining projects – a shift that could be associated with broader developments in the anthropology of capitalism (see Bear et al. 2015).

This last line of analysis provides fertile ground for empirical investigation in the future. There is still very little anthropological research on how mining companies' labour policies are put in place, their consequences for mineworkers, and the broader social dynamics they generate. Beyond labour policies, it is necessary to broaden the research scope to include new categories of mine employees and to pay more attention to the content of their work. So far, anthropologists have favoured the study of workers and, to a lesser extent, expatriate executives; some have also provided insights into the work of employees in the social departments. Even though they play a key role in the way mining projects are developed, research on middle managers and technicians – especially those working in departments related to exploration, production, maintenance, and finance – is almost non-existent.

Furthermore, researchers' interest has focused more on conditions of employment than on the content of their work. Following Rolston (2013; see also Ferry 2005), more attention should be paid to what workers do – their concrete tasks, their interactions with machines, their use of documents, and more generally, the way they transform minerals into commodities (Richardson and Weszkalnys 2014). Such a line of research would allow the study, in more detail, from an insider's view, of various processes that are attracting growing interest in the field of mining studies: the changing geology of mining, its environmental impact, the mechanization of production, and the financialization of the industry.

From a comparative perspective, there are few attempts at researching the various forms of work involved in mining, their evolution, their diversity, and their interrelations (see, for example, D'Angelo 2018). This broad research agenda would involve going beyond the focus on greenfield mining projects funded by large corporations headquartered in the Global North, to comparing different types of mining projects depending on the mineral that is mined, the type of mine being operated, its location, and the parent company's identity (Lee 2017, 2018; Rubbers 2021). It also implies connecting the work of industrial and artisanal miners into a broader reflection on the articulation between modes of production, the dynamics of labour markets, and the integration of mines in global value chains (Rubbers 2019; Verbrugge and Geenen 2020). Although labour occupies a significant place in the anthropology of artisanal mining, it is often approached in this literature as a distinct domain of practices and representations. As a number of scholars remind us, however, artisanal mining is deeply integrated into transnational value chains, and this since pre-colonial times (Herbert 1984; Dumett 1998; Nixon 2017). Thus, the livelihood opportunities it provides cannot be entirely disconnected from labour market dynamics and the global political economy. Finally, it is necessary, in order to better understand the dissemination of production, surveillance, and labour management techniques on a global scale, to de-compartmentalize research from regional traditions and area studies.

Such a perspective would allow research to go beyond the approach prevailing in the public debate on mining and development (see also Pijpers this volume), which tends to think of work only in terms of the number of jobs created and income levels. To be sure, such a quantitative approach is politically important in order to measure, rank, and compare the contribution made by the mining industry in different countries. However, it does not allow us to understand how mining projects work. An ethnographic approach centred on everyday work practices and the power relations they involve remains, in this respect, irreplaceable. Although it has become marginal since the early 2000s, such an approach still has a lot to teach us about the social life of mining projects, their effects on the countries where they are established, and the dynamics of mining capitalism.

Notes

1 For reasons of space and consistency, this review focuses on the anthropology of industrial mineworkers, that is, the employees of industrial mining companies (companies using mechanical means to extract and process the ore). It does not cover historical and sociological studies on the subject, nor the related, yet distinct, body of literature on artisanal mining.

2 A recent article by Thomas McNamara and Robby Kapesea (2020) on trade unions in the new mining projects of the Northwestern province in Zambia suggests that Epstein's analysis of the ambiguities between tribal and union loyalties still has some relevance.

3 Nash's approach to mineworkers' culture seems inspired by cultural Marxism, especially Edward Thompson (1963). This influence remains, however, largely implicit in the book.

4 Like most studies of working-class culture in the 1970s and 1980s, *We Eat the Mines and the Mines Eat Us* could be criticized for re-inscribing the particularities of Bolivian

workers' traditions into a universal metanarrative, that of their proletarianization and the making of a militant working class (see Cooper 1995; Sewell 1993).

5 This analytical shift has been influential for my own *Le paternalisme en question* (Rubbers 2013). In this book, the paternalistic policy put in place by Union minière, and then by Gécamines, in the Congolese Copperbelt is interpreted as having contributed to structure workers' subjectivity, the sense that they have of themselves as workers, husbands, fathers, or citizens.

6 This collective and comparative research was carried out within the framework of the WORKinMINING project (see www.workinmining.ulg.ac.be).

Bibliography

Ballard C. and Banks G. (2003) Resource wars: The anthropology of mining. *Annual Review of Anthropology* 32: 287–313.

Bear L., Ho K., Tsing A.L.Yanagisako S. (2015) Gens: A feminist manifesto for the study of capitalism. Theorizing the contemporary, *Fieldsights* (30 March). https://culanth.org/field sights/gens-a-feminist-manifesto-for-the-study-of-capitalism.

Carsten J. (2004) *After kinship*, Cambridge: Cambridge University Press.

Cooper F. (1995) Work, class and empire: An African historian's retrospective on E.P. Thompson. *Social History* 20 (2): 235–241.

Cooper F. (1996) *Decolonization and African society. The labor question in French and British Africa*, Cambridge: Cambridge University Press.

D'Angelo L. (2018) From traces to carpets: Unravelling labour practices in the mines of Sierra Leone. In: De Vito C. and Gerritsen A. (eds) *Micro-spatial histories of global labour*. Cham: Palgrave-MacMillan, 313–342.

D'Angelo L. and Pijpers R.J. (2018) Mining temporalities: An overview. *Extractive Industries and Society* 5 (2): 215–222.

Dumett R E. (1998) *El Dorado in West Africa: The gold-mining frontier, African labor, and colonial capitalism in the Gold Coast, 1875–1900*. Oxford: James Currey Publishers.

Epstein A.L. (1958) *Politics in an urban African community*, Manchester: Manchester University Press.

Epstein A.L. (1981) *Urbanization and kinship: The domestic domain on the Copperbelt of Zambia, 1950–1956*, New York: Academic Press.

Evens T.M. and Handelman D. (2006) *The Manchester School. Practice and ethnographic praxis in anthropology*, Oxford: Berghahn.

Ferguson J. (1990a) Mobile workers, modernist narratives: A critique of the historiography of transition on the Zambian Copperbelt [part one]. *Journal of Southern African Studies* 16 (3): 385–412.

Ferguson J. (1990b) Mobile workers, modernist narratives: A critique of the historiography of transition on the Zambian Copperbelt [part two]. *Journal of Southern African Studies* 16 (4): 603–621.

Ferguson J. (1994) Modernist narratives, conventional wisdoms, and colonial liberalism: Reply to a Strawman. *Journal of Southern African Studies* 20 (4): 633–640.

Ferguson J. (1999) *Expectations of modernity. Myths and meanings of urban life on the Zambian Copperbelt*, Berkeley: University of California Press.

Ferry E. (2005) *Not ours alone. Patrimony, value, and collectivity in contemporary Mexico*, New York: Columbia University Press.

Ferry E. and Limbert E. (2008) *Timely assets: The politics of resources and their temporalities*, Santa Fe: School for Advanced Research Press.

Filer C. and Le Meur P.Y. (2017) *Large-scale mines and local-level politics. Between New Caledonia and Papua New Guinea*, Acton: Australian National University Press.

Finn J.L. (1998) *Tracing the veins: Of copper, culture and community from Butte to Chuquicamata*, Berkeley: University of California Press.

Geenen K. (2020) Patronage, social proximity, and instrumentality in the mining industry in the Democratic Republic of Congo: The union elections explored. *Dialectical Anthropology* 44: 173–185.

Geertz, C. (1973) *The interpretation of cultures. Selected essays*, New York: Basic Books.

Gluckman M. (1945) The seven year research plan of the Rhodes-Livingstone Institute, human problems in British Central Africa. *Rhodes-Livingstone Journal*: 1–32.

Goffman E. (1990 [1959]) *The presentation of self in everyday life*, New York: Penguin.

Golub A. (2014) *Leviathans at the gold mine. Creating indigenous and corporate actors in Papua New Guinea*, Durham: Duke University Press.

Gordon, R.J. (1977) *Mines, masters and migrants. Life in a Namibian mine compound*, Johannesburg: Ravan Press.

Hansen T.H. (1991) After *Copper Town*: The past in the present of urban Zambia. *Journal of Anthropological Research*. 47 (4): 441–456.

Harvey D. (2003) *The new imperialism*, Oxford: Oxford University Press.

Herbert E.W. (1984) *Red gold of Africa. Copper in precolonial history and culture*. Madison: University of Wisconsin Press.

Hier S. and Kemp C. (2002) Anthropological stranger: The intellectual trajectory of Hortense Powdermarker. *Women's History Review* 11 (2): 253–272.

James D. and Rajak D. (2014) Credit apartheid, migrants, mines and money. *African Studies* 73 (3): 455–476.

Kapesea R. and McNamara T. (2020) "We are not just a union, we are a family": Class, kinship and tribe in Zambia's mining unions. *Dialectical Anthropology* 44 (2): 153–172.

Kesküla E. (2016) Temporalities, time and the everyday: New technology as a marker of change in an Estonian mine. *History and Anthropology* 27 (5): 521–535.

Kesküla E. (2018) Risky encounters: The ritual prevention of accidents in the coal mines of Kazakhstan. In: Pijpers R.J. and Eriksen T.H. (eds) *Mining encounters. Extractive industries in an overheated world*. London: Pluto Press, 156–173.

Kesküla E. and Sanchez A. (2019) Everyday barricades: Bureaucracy and the affect of struggle in trade unions. *Dialectical Anthropology* 43: 109–125.

Kirsch S. (2014) *Mining capitalism. The relationship between corporations and their critics*, Oakland: University of California Press.

Larmer, M. (2007) *Mineworkers in Zambia. Labour and political change in post-colonial Africa*, London: I.B. Tauris.

Lee C.K. (2017) *The spectre of global China: Politics, labor, and foreign investment in Africa*, Chicago: Chicago University Press.

Lee C.K. (2018) Varieties of capital, fracture of labor. A comparative ethnography of subcontracting and labor precarity on the Zambian Copperbelt. In: Hann C. and Parry J. (eds) *Industrial labor on the margins of capitalism. Precarity, class, and the neoliberal subject*, Oxford: Berghahn, 39–60.

Li F. (2015) *Unearthing conflict. Corporate mining, activism, and expertise in Peru*, Durham: Duke University Press.

Li T.M. (2009) To make live or let die? Rural dispossession and the protection of surplus populations. *Antipode* 41 (1): 66–93.

Macmillan H. (1993) The historiography of transition on the Zambian Copperbelt – Another view. *Journal of Southern African Studies* 19 (4): 681–712.

Macmillan H. (1996) More thoughts on the historiography of transition on the Zambian Copperbelt. *Journal of Southern African Studies* 22 (2): 309–312.

Marx K. (1887) *Capital. A critique of political economy. Volume 1: The process of production of capital*, Moscow: Progress Publishers.

McNamara T. (2021a) A reasonable negotiation? Workplace-based unionists' subjectivities, wage negotiations and the day-to-day life of an ethical-political project. *Journal of the Royal Anthropological Institute* 27 (3): 617–637.

McNamara T. (2021b) The union has reoriented towards entrepreneurship: Neoliberal solidarities on Zambia's Copperbelt. *Third World Quarterly* 42 (9): 2152–2171.

Musonda J. (2020) Undermining gender: Women mineworkers at the rockface in a Zambian underground mine. *Anthropology Southern Africa* 43 (1): 32–42.

Musonda J. (2021) Modernity on credit: The experience of underground miners on the Zambian Copperbelt. *Journal of Southern African Studies* 47 (3): 369–385.

Nash J. (1993 [1979]) *We eat the mines and the mines eat us. Dependency and exploitation in Bolivian tin mines*, New York: Columbia University Press.

Nixon S. (2017) Trans-Saharan gold trade in pre-modern times: Available evidence and research agendas. In Mattingly D., Leitch V., Duckworth C., Cuénod A., Sterry M. and Cole F. (eds) *Trade in the ancient Sahara and beyond*. Cambridge: Cambridge University Press, 156–188.

Powdermaker H. (1962) *Copper town: Changing Africa*, New York: Harper and Row.

Pugliese F. (2020) Mining companies and gender(ed) policies: The women of the Congolese Copperbelt, past and present. *Extractive Industries and Society*. Available at: https://www.sciencedirect.com/science/article/pii/S2214790X2030246X.

Rajak D. (2011) *In good company. An anatomy of corporate social responsibility*, Stanford: Stanford University Press.

Rajak D. (2016) Hope and betrayal on the platinum belt: Responsibility, violence and corporate power in South Africa. *Journal of Southern African Studies* 42 (5): 929–946.

Richardson T. and Weszkalnys G. (2014) Introduction: Resource materialities. *Anthropological Quarterly* 87 (1): 5–30.

Rolston J.S. (2010) Risky business: Neoliberalism and workplace safety in Wyoming coal mines. *Human Organization* 69 (4): 331–342.

Rolston J.S. (2013) The politics of pits and the materiality of mine labor: Making natural resources in the American West. *American Anthropologist* 115 (4): 582–594.

Rolston J.S. (2014) *Mining coal and undermining gender. Rythms of work and family in the American West*, New Brunswick: Rutgers University Press.

Rubbers B. (2013) *Le paternalisme en question. Les anciens ouvriers de la Gécamines face à la libéralisation du secteur minier katangais (R.D. Congo)*, Paris: L'Harmattan.

Rubbers B. (2019) Mining boom, labour market segmentation and social inequality in the Congolese Copperbelt. *Development and Change* 51 (6): 1555–1578.

Rubbers B. (2020a) Governing new mining projects. The view from the HR department of a Chinese company in the Congolese Copperbelt. *Extractive Industries and Society* 7 (1): 191–198.

Rubbers B. (2020b) Company brokers: Human resources managers in foreign mining projects in the Congolese Copperbelt. *Ethnography*. Available at: https://journals.sagepub.com/doi/pdf/10.1177/1466138120948911.

Rubbers, B. (2021) *Inside mining capitalism. The micropolitics of work in the Congolese and Zambian Copperbelts*. Oxford: James Currey.

Schumaker L. (2001) *Africanizing anthropology. Fieldwork, networks, and the making of cultural knowledge in Central Africa*, Durham: Duke University Press.

Scott J. (1990) *Domination and the arts of resistance. Hidden transcripts*, New Haven: Yale University Press.

Sewell W.H. (1993) Towards a post-materialist rhetoric for labor history. In: Berlanstein L. (ed.) *Rethinking labor history: Essays on discourse and class analysis*. Urbana: University of Illinois Press, 15–38.

Shever E. (2012) *Resources for reform. Oil and neoliberalism in Argentina*, Stanford: Stanford University Press.

Thompson E. (1963) *The making of the English working class*, New York: Pantheon.

Verbrugge B. and Geenen S. (2020) *Global gold production touching ground. Expansion, informalization, and technological innovation*, Cham: Palgrave-MacMillan.

Welker M. (2014) *Enacting the corporation. An American mining firm in post-authoritarian Indonesia*, Berkeley: University of California Press.

Werbner R.P. (1984) The Manchester School in South-Central Africa. *Annual Review of Anthropology*. 13: 157–185.

8

POSITIONALITY AND ETHICS

Nicholas Bainton and Emilia E. Skrzypek

1 Introduction

We are now living in the "mineral age" (Jacka 2018) where the global economy, and our daily lives, are thoroughly entwined with the perpetual extraction and exploitation of natural resources. Making sense of these resource encounters and relations requires creative ways of researching extractive processes, and methods capable of engaging diverse sets of actors and interests across contested terrains. Different situations may require different positional stances, which in turn may require different methodological approaches. In this chapter, we summarize some of these positions and the ethical implications of these ethnographic engagements in resource extraction contexts, recognizing that new positions are continually emerging, these positions can overlap, and that many anthropologists move between them over the course of their career as they investigate extractive effects. We are primarily concerned with large-scale mining, and oil and gas extraction. While many of our observations are also relevant for researchers working in arti-sanal and/or small-scale mining contexts (e.g. Bainton et al. 2020; Luning 2014), we focus here on large-scale projects where power asymmetries and the production of harm for profit are especially pronounced, adding salience to questions of research ethics and positionality. We conclude by charting new directions for the ethics of an anthropology of resource extraction.

2 Defining the field of resource extraction

Ever since Bronislaw Malinowski set a new standard for the ethnographic method – a century ago (Malinowski 1961 [1922]) – the "field site" took centre place in anthropological research and analysis, and immersive research in "the field" based on long-term participant observation emerged as the defining method of the

DOI: 10.4324/9781003018018-8

discipline and the main factor determining research as "anthropological" (Gupta and Ferguson 1997). Early anthropological works on mining generally reflected this approach and focussed on mines and mining communities as somewhat bounded sites of ethnographic enquiry (e.g. Gordon 1977; Harris 1959). As the discipline evolved, it was challenged to expand from what George Marcus (1995) called the "committed localism" of early anthropology and rethink the definition of a "field site" and the content of the disciplinary toolkit (Coleman and Collins 2006). Marcus proposed a shift from single-sited ethnography which saw field sites as neatly defined objects of study to a "multi-sited" approach where "ethnography of a cultural formation in the world system is also an ethnography of the system" (1995: 99). Applied to the context of natural resource extraction, this approach contends that the anthropology of resource extraction is, or should be, a study both *in* and *of* resource settings and globalized extractive capitalism (e.g. Tsing 2005). What emerges are messy, contentious and dynamic research contexts – what Pierre-Yves Le Meur (2015) terms "resource arenas" – comprised of multiple actors and interfaces, shaped by local and global forces.

This view of unbounded extractivism has been conceptualized by some Brazilian anthropologists in terms of "spillage effects" (Gudynas 2016 cited in Zhouri 2017). Conjuring images of industrial overflow (Letté 2009) or creeping impacts, attention is directed to those extractive processes that occur across time and space beyond the point of extraction, for example in global institutions (Sawyer and Gomez 2012); state and corporate domains (Bainton and Skrzypek 2021); infrastructure (Richardson and Weszkalnys 2014; Szeman 2017); energy, economic, and legal systems (Love and Isenhour 2016; Smith and High 2017); corporate discourse (Dolan and Rajak 2016) and beyond. These new research spaces have driven the anthropology of extraction beyond a traditional focus on alienated labourers and marginalized communities – focal points that defined this field of inquiry in the classic works of June Nash (1979) and Michael Taussig (1980) – and broadened the scope of ethnographic inquiry to include a diverse set of actors, assemblages, institutions, and processes.

The contemporary literature on the anthropology of extraction at a large scale can be grouped into two broad categories: ethnographic works on the effects of resource companies and extractive capitalism (e.g. Bainton 2010; High 2017; Li 2015) and ethnographic works on the internal dynamics of resource companies, their extractive projects and employees (e.g. Carstens 2001; Dalsgaard 2019; Smith 2019). There is of course some overlap between these categories, but for the most part, the anthropological literature on large-scale resource extraction tends to fall within the former. This is partly due to the prevailing disciplinary interest in the lives of subaltern others, and what Ricardo Godoy once described as "the anthropologists' penchant for, and strengths with, small-scale social systems" (1985: 211). Saying that, many anthropologists have heeded Laura Nader's (1974, 1997) call to "study up" and turn their attention to contemporary institutions of power. Exemplary ethnographies include Merina Welker's (2014) work on the operations of an American mining firm in Indonesia; Dinah Rajak's (2011) ethnography of an international mining corporation's social responsibility programmes and operations in South Africa; and Jessica Rolston's (2014) exploration of the lives of coal miners in the United States.

In the contentious and dynamic contexts of large-scale resource extraction, "studying up" and analysing what Nader (1974, 1997) termed "dominant institutions" or what she would later call "controlling processes" is one of many possible and, indeed, necessary directions for researching "extractive relations" (Owen and Kemp 2017). It is important to note, however, that even the "studying up", as envisaged by Nader, actually entails studying down and sideways to connect issues and happenings emergent on the ground to decisions and actions of powerful institutions like multinational corporations head quartered in major financial districts (Stryker and González 2014) and further to global socio-economic processes like share markets or shareholder activism (Welker and Wood 2011). There is also an important temporal dimension to the anthropology of extraction – one that builds on the discipline's ability to work with local histories and study societies and phenomena in the present, and that often extends to the future through the prism of local aspirations or extractive legacies (Voyles 2015), or in some cases, the production of knowledge to inform future orientated decisions (see Ferry and Limbert 2008 on timely assets; and D'Angelo and Pijpers 2018 on mining temporalities). To that end, Ulf Hannerz observes, "anthropologists now study not only here or there – through or sideways – but also backward and forward" (2010: 60).

Defining the field and locating the anthropologist in this field leads to fundamental questions of gaining access to the field. An anthropologist's ability (or, as the case might be, inability) to secure access to a given site is influenced by a range of factors linked to their personal and institutional research interests and agendas, and their capacity to access and mobilize social, cultural, institutional, and financial capital (see Straube 2020 for a perspective from Zambia's Copperbelt) – all of which entail a range of practical and ethical considerations. In many cases, access to communities in mining localities involves little by way of formalized research agreements and approvals, the terms of the relationship being typically subject to ongoing negotiation through a form of social contract – a series of interpersonal or community expectations and obligations (e.g. Jacka 2015). Similarly, anthropologists usually encounter fewer difficulties accessing company employees who work in "community-facing" roles (e.g. Skrzypek 2020), or employees or lobbyists who are also local community residents (e.g. Eriksen 2018). However, gaining access to the most powerful people, including operational level general managers and corporate level executives, can be more problematic and may entail negotiating formal access agreements, or confidentiality agreements. In these private organizations, the combination of bureaucratic requirements, suspicion, internal political machinations, and the commercial, legal, and political sensitivities that surround their operations can conspire against easy ethnographic access. Of course, these same observations are often true for local communities, and we do not presume that community members are necessarily keener than corporate executives to be studied. However, the power differences between these actors can result in rather different motivations for opening or blocking ethnographic access.

Whilst some anthropologists get caught up in resource encounters when extractive companies enter their already existing research locations (e.g. Minnegal and Dwyer 2017), others find themselves negotiating access to extractive processes with actors who can serve as a gateway to the field site (e.g. Lee 2009). In either case, access is not something that is negotiated once and then forgotten, nor is it something that is negotiated with only one person or a narrowly defined group. Rather, it is something that is continually negotiated throughout the time of fieldwork (and often beyond) with a broad range of actors. It is therefore inherently political and ethical in nature (Cunliffe and Alcadipani 2016), and we do well to scrutinize the ways that access and its influence on the position of anthropologists in and in relation to the field transforms research and informs the production of knowledge – as we do below.

3 Positionality in the field

The anthropology of resource extraction has its roots in the ethnographic work conducted under the auspices of the Rhodes-Livingstone Institute (RLI) in what was formerly British Central Africa (see Bainton 2020). The RLI anthropologists, also known as the "Manchester School" because of their connection with Manchester University, produced a number of classic essays and monographs – later dubbed the "Copperbelt studies" – that focussed on the various socio-economic processes that accompanied the development of industrial mining activities across the region (e.g. Epstein 1958; Gluckman 1961; Mitchell 1959; see also Ferguson 1999). While the RLI anthropologists saw themselves as progressives and advocates for African interests, some post-colonial scholars argued that they failed to address the legacies of colonial oppression – they were, in the end, part of the colonial establishment, not an alternative to it (Magubane 1971). These critiques foreshadowed contemporary debates about ethnographic positionality vis-à-vis extractive companies and other powerful actors. Such debates generally pivot on the politics of engagement in power-laden fields, and they parallel other well documented debates concerning the relationship between anthropology and advocacy (Goodale 2006; Abu-Lughod 1990); the role of applied anthropologists (Stewart and Strathern 2005); or the division between so-called "pure" and "profane" forms of anthropology which locates academic anthropology "in a 'sacred' space that is distinguishable from applied research occupying a less pure intellectually inferior and more morally profane domain" (Trigger 2011: 233).

Debates about the positionality of anthropologists studying the extractive industries were later rearticulated with great force in the Melanesian context. In the 1990s, a good number of anthropologists and other social scientists found themselves entangled in the triangular relationship between extractive companies, local communities, and government agencies in Papua New Guinea (e.g. Burton 1999; Filer 1999; Hyndman 1995; Jackson 1991). This is partly because Papua New

Guinea has long been a source of ethnographic attention, but also because unlike some other post-colonial settings, especially throughout Africa, there had not been a political reaction against the discipline after Papua New Guinea gained independence in 1975. Some resident anthropologists were involved in formulating national laws and policies relating to the extractives sector, while those based overseas could obtain permits to conduct ethnographic research in communities affected by extractive projects and publish findings that were critical of the government, extractive companies, or both (Filer 2018: 6). If large-scale resource extraction projects in Papua New Guinea were now subject to greater ethnographic scrutiny than their counterparts in other countries, there was also very little consensus among the anthropologists who worked there as to the position that they should assume within these arenas of extraction. This led Chris Ballard and Glenn Banks to conclude that "there is no natural position for anthropology in such a contested field" (2003: 306).

These differences of opinion over the role of anthropologists were possibly best articulated in a series of scholarly exchanges between Stuart Kirsch and Colin Filer over the merits and pitfalls of advocacy and activism. Kirsch had first undertaken ethnographic research with the communities living downstream from the environmentally disastrous Ok Tedi mine in Papua New Guinea, and later found himself entangled in their legal struggles against the operator of the mine. This led him to argue that anthropological activism is a natural extension of the reciprocity that defines the ethnographic pact, and that in order to shift the balance of power, anthropologists should form robust global alliances with Indigenous people, lawyers, and NGOs (Kirsch 1996, 2002). Having played a pivotal role in advancing the opportunities to apply anthropology in resource extraction settings in Papua New Guinea, Filer argued that anthropologists are better placed to play the role of "honest broker", assisting the different actors involved in these arenas to understand their respective expectations and points of view (Filer 1999: 89). More than a decade later Alex Golub and Mooweon Rhee attempted to reconcile these positions by suggesting that anthropologists could instead position themselves in extractive arenas as a kind of "ombudsman" (Golub and Rhee 2013). This role would avoid the risks of partisanship and allow anthropologists to understand and publically express the views of those on the other side of the corporate divide even if they disagree with their actions and opinions.

These debates highlight the interdependence between the ways in which anthropologists conceptualize the field and think about resource arenas, the positional stances they may choose to occupy, and the type of knowledge they may produce. For example, Nick Bainton and John Owen have proposed that we can think of resource arenas as "zones of entanglement" that entail a multiplicity of interfaces, actors, and networks characterized by "ongoing contestation and negotiation over meaning, values and interests" – an approach that calls researchers "to embrace the interests of all actors involved in the making, unmaking and remaking of mining scenarios and their outcomes" (2019: 769). Others have conceptualized resource arenas through a prism of severe structural power inequalities and used their position to redress the imbalance and advocate on behalf of communities, or support their acts of resistance against extractive industries (e.g. Kirsch 2014). This type of "engaged anthropology" can involve performing the role of expert witness (Rosen 1977), writing affidavits, collaborating in peace campaigns, and other forms of activist involvement in political and legal processes. The value of this approach is the emphasis it can place on community interests, and the potential for developing innovative ethnographic methods and having an impact "beyond the text" (Kirsch 2018). If this type of framing prompts some anthropologists to argue that the only reasonable and ethical course of action is a form of activist-anthropology that is personally engaged and politically committed (e.g. Jalbert et al. 2017), it can also help draw attention to the violence of extractive capitalism and the agency expressed through subaltern struggles (Sawyer 2004). A well-known example of this approach in the Amazon context is the work of the ethno-ecologist, Darrell Posey, who "felt compelled to participate in the Kayapo's battles to protect their rainforest home" (Girardet 2001).

Interpreted within the polarized view of community–corporate relations and using Robert Foster's (2017) classification of anthropological approaches to the study of corporations, the pro-community position also tends to be an anti-corporation position. In some cases, this can lead to a form of "ethnographic refusal" (Ortner 1995), where deliberate decisions are made to avoid direct engagement with company actors in order to preserve anthropological allegiances to affected groups – or to avoid the risk of "pollution" (Douglas 1966). However, some anthropologists choose to collaborate with the wider set of "stakeholders" who are entangled in these extractive settings in an attempt to facilitate beneficial outcomes for local people and the environment. These forms of collaboration usually result in alliances that extend beyond the triumvirate of communities, lawyers, and NGOs to encompass other actors and institutions including state regulators, policy makers, or international finance institutions, as well as extractive companies themselves. This can occur through new research consortiums and networks (Le Meur 2015), or forms of applied anthropology or contracted research, including social impact assessment (Goldman 2000), "social mapping" and landowner identification studies (Weiner 2002), cultural heritage management (Bainton et al. 2011), and other forms of professional advice. The advantage of this type of work is that it can provide anthropologists with access to places, people, and events that may otherwise be difficult to achieve, and it often results in ideas that transcend the initial research agenda or scope. On the other hand, these forms of applied engagement are subject to significant scrutiny on the grounds of power imbalances and vested interests. Critics argue that any professional engagement that ultimately enables extractive projects to be licenced or to continue operating – which inevitably leads to social and environmental harm – represents a form of corporate capture (e.g. Coumans 2011; but see also Aiello 2012; and Burton 2014). Assessed from this perspective, anthropologists have a moral obligation to oppose these forms of capitalist expansion by whatever means possible. When extractive companies commission, fund, and utilize this work to advance their own objectives, this position can be described as "instrumentalist", in that it regards anthropology as a means to an end, where anthropologists are active participants "within" these extractive encounters (Golub 2018: 23) and are therefore complicit (Marcus 1997) in the effects of extractive processes and relations, as we discuss below.

The instrumentalist argument gains further traction in those situations where anthropologists work in corporate structures. The proliferation of international standards for corporate social responsibility (Dashwood 2012) has provided ample opportunities for anthropologists to ply their trade within the broader extractives sector to help promote these standards, and to assist companies and governments to manage the "social risks" associated with these industrial investments. Whilst some anthropologists perform these roles with NGOs (e.g. Coumans 2017, 2018) others work directly for resource companies. They are mostly employed in community-facing roles, or what the industry prefers to call the "social performance" function, which can include responsibility for community relations and community development programmes, and more specific tasks such as social impact monitoring programmes, cultural heritage management, or resettlement, among others (Kemp and Owen 2018). Notwithstanding other professional motivations, some corporate anthropologists claim they are better positioned to advance social and environmental interests because they can influence decision making processes and the allocation of internal resources – what they might call "making the business case" for social responsibility (e.g. Cochrane 2017).

Some of these embedded anthropologists carry out research activities. In most cases, however, they perform technical and managerial roles, and their efforts are mainly directed towards specific corporate and operational goals. If these anthropologists are better placed to observe the messy entanglements within resource arenas, their research is often proprietary, or their knowledge remains confidential, and they are highly constrained in their ability to publicly advance radical critique. They typically occupy a non-academic, *professional* position in the workplace as a researcher, consultant, manager, or designer. As Melissa Cefkin observed in her work on corporate ethnography, in these settings "the anthropologist operates as a mutual corporate actor with other members of the corporation" (2012: 95).

By contrast, anthropologists who conduct research inside and under the auspices of a corporation but are employed outside the corporation (i.e. at a research institution), are potentially better positioned to offer more critical perspectives on company operations and community dynamics, and may enjoy a greater ability to move between different actors and positions. But, as we have noted above, these researchers also face challenges gaining access to the inner workings of resource companies, and they may have to convince their managerial gatekeepers that their research will be beneficial for the company in exchange for access. Which underscores the point that access is always contingent upon *interest* – just as with local communities, where anthropologists must find someone who *feels* interested in their work and negotiate relations of obligation, someone within the corporation, government, or NGO must feel that this is a worthwhile research idea and get behind the project. This was certainly the case for Marina Welker (2014) who relied upon the curiosity and intellectual interest of Newmont's social responsibility executives – one of whom happened to be an anthropologist – in order to access their corporate headquarters and their Indonesian operation.

4 Implications for methods and knowledge

These positional stances may require different methodological approaches and may sometimes result in a bias towards certain methods. Participant observation is essential content for any anthropological toolbox. The type of immersive fieldwork associated with anthropology facilitates in depth, detailed research, and for many it constitutes the unique value of anthropology. But as anthropologists who have tried to "study up" can attest (e.g. Nugent and Shore 2002), participant observation does not easily travel up the social ladder and applying this approach to the inner workings of resource extraction companies is not always a feasible proposition. In those settings, the classic ideal of participant observation as "anthropology by immersion" often resembles a form of "anthropology by appointment" (Hannerz 2010: 77) – where access to interlocutors and interactions in the field "are often limited, regulated, and timed" and may include pre-arranged, structured interviews, informal conversations with government and company personnel outside their work setting, and analysis of secondary data (see Espig and de Rijke 2018).

To circumnavigate these obstacles, innovative approaches to studying mining, oil, and gas companies have developed within the discipline. Some anthropologists have turned their focus to the discourse and practice of corporate social responsibility (Idemudia 2010; Kapelus 2002), combining ethnographic methods with discourse analysis. Others have adopted a version of what Sherry Ortner (2010) terms "interface ethnography" – which shifts the focus to the ways that closed institutions, like resource extraction companies, interface with the public. This includes, for instance, direct engagements between company representatives and communities; their external communications; their appropriation and manipulation of critique (Bensen and Kirsch 2010); and how they present themselves in project locations, or how they are "enacted" (e.g. Golub 2014). For anthropologists who undertake contracted research for extractive companies or governments, and embedded anthropologists, terms of engagement are often defined in contractual obligations and confidentiality agreements placing restrictions on research activities as well as the ownership and dissemination of data. In those contexts, the choice of methodology is conditioned, in part at least, by the kind of work required by the client and influenced by time, resources, and operational factors. Structured interviews, focus groups and different kinds of survey have all been used in these circumstances, often driven by the need for pre-defined kinds of data about the social context and the consequences of and for specific operations (e.g. Banks 2013; de Rijke 2013).

The methodological implications of particular positionalities extend to the types of knowledge being produced, how it is communicated and circulated, and to whom. One of the main arguments for anthropology's involvement in resource extraction contexts stems from its perceived ability to mediate between the different epistemologies that populate sites of extraction (e.g. Gilberthorpe 2013; Weszkalnys 2010). In the second part of the twentieth century this disciplinary trait was commonly known as "cultural translation" (Asad 1986). And while the concept has lost its popularity in recent decades, as the "crisis of representation" and debates about "writing culture" in the 1980s and 1990s questioned "the empirical possibility of understanding others in their own terms" (Jackson 2013: xi), the role of anthropologists in extractive contexts continues to be premised on the ability to work with or broker the different lifeworlds occupying resource arenas. Local epistemologies do not easily translate into "data" or "information points" of the kind sought after by extractive companies or governments, just as the technical or bureaucratic language employed by companies and governments is not always easily translated into local epistemologies. Nevertheless, most contracted research is based on the assumption that enhancing mutual understanding between the different actors in resource contexts, especially those who occupy positions within the company or the community, can help to reduce social and environmental risks and impacts.

Critics of output-driven approaches employed by companies or governments argue that short-term, targeted research engagements prohibit the type of in-depth understanding that comes with immersive ethnographic fieldwork that, they argue, grants the ethnographer a privileged access to knowledge and relations (e.g. Macdonald 2002). In this view, corporate or legal modes of knowledge production are not able to grasp and accommodate the intricacies of local epistemologies, instead entailing an act of simplification which serves corporate or state ends over the needs and goals of local communities. However, there are also cases where communities have sought out and employed new ways of presenting (and simplifying) their knowledge and their claims in ways which they perceive as being more efficacious in resource contests (e.g. Bell 2014; Dwyer and Minnegal 2014). In some instances, communities may even seek out the services of anthropologists to help them with this task (Burton 2007), or to access anthropological materials to support their claims. Here access is multi-directional. More broadly, there has been a concerted effort within anthropology (partly resulting from the requirements of funding agencies) to ensure that research outputs are accessible to communities – in terms of physical (or virtual) access to material outputs (e.g. books), but also their readability and value for community members.

This brings us to an important point. While the concerns and methodologies of anthropologists conducting contract research may be influenced by the needs of the company, government, or NGO paying for their services – the same can be said for anthropologists who position themselves as advocates for local communities in their fight for recognition of their rights to land. Just as contract and embedded anthropologists may tailor their results to the needs of their employers, activist-anthropologists may also choose or be forced to temper their accounts of extractive relations to meet the expectations of local communities, or to ensure that they do not undermine the political cause of those with whom they are aligned. Further, as part of their work contracts, consultants and embedded anthropologists nearly always face restrictions on whether they can make the results of their research publically available, at what time, and under what conditions. Indeed, these restrictions are one of the main arguments used by critics to question the ethics of applied, engaged, and/or embedded anthropology in extractive settings. Although it is rare for anthropologists to sign formal research agreements with local communities, the social contracts they negotiate with members of those communities may also influence the form and content of research outputs – and have ethical implications.

5 Where to next for the ethics of anthropological encounters with resource extraction?

However we interpret them, resource arenas are complex and contested spaces and Ballard and Banks' earlier observation that there is no automatic position for anthropology in extractive spaces remains just as true today as it was nearly 20 years ago. If anything, in that time, the growing recognition of Indigenous rights, combined with the pressure on the industry to acknowledge and address the social impacts of their operations, have created new demands and new opportunities for anthropological engagements with and in resource arenas. But if the broadening of the field means that anthropologists are now even more entangled in those spaces, how do they navigate the ethical dimensions of these encounters?

Mitchell Sedgwick (2017: 62) writes that "across the discipline's history, as today, the vast majority of anthropologist fieldworkers have empathized genuinely with the circumstances of their host communities" – and we find it impossible to disagree. The question that has emerged within debates about the anthropology of extraction is whether such genuine empathy can, or should be extended to other agents, such as company employees, in line with what Thomas Kirsch (2007: 98) glosses as a "principled symmetry" that "stresses the researcher's moral responsibility and socio-political accountability" towards all interlocutors. These sentiments are echoed in Mette High and Jessica Smith's (2019) recent call for a more capacious methodology that allows for ethnographic exploration and understanding of the "ethical worlds" of different actors in resource arenas. While this stresses a certain fidelity to a holistic ethnographic method, we might also ask whether and/or how "principled symmetry" is a feasible and, indeed, an ethical proposition in contexts commonly characterized by asymmetries of power and crude demonstrations of socio-political inequalities. And, considering the issues surrounding positionality and ethics, how might anthropologists "assess their responsibility for structurally unjust mining encounters" (Golub 2018: 29)?

To some extent, the answer to that question is both a matter of personal ethics and research interests. In the 1990s and early 2000s, a number of anthropologists openly debated the ethics of different positional stances assessed predominantly on the grounds of personal choices and value judgements – as witnessed in the debates between Stuart Kirsch and Colin Filer (see also Strathern and Stewart 2001). The "crisis of representation" which began in the 1970s, and the conversations about "writing culture" and anthropology's reflexive turn which followed in the 1980s (Clifford and Marcus 1986) led some anthropologists to critically consider questions of ownership of anthropological data and knowledge (Geer 2003); complex regimes of accountability in which anthropologists operate (Morreira 2012); opportunities for co-labouring with interlocutors (De la Cadena 2015); and the ethics of anthropologists speaking for Indigenous communities in resource contexts (Wilson and Swiderska 2009). Elaborating on this final point, Smith and High write that scholars ought to be "self-reflexive about their own political commitments, to recognize when and how these influence our interpretation of ethnographic evidence, and to open up intellectual space for our interlocutors to enact and express commitments that differ from our own" (2017: 5).

There seems to be consensus that, regardless of their positional stance, anthropologists are responsible for the effects of their work and engagements, and the relationships and networks they have shaped and mobilized in order to conduct their work (Golub 2018). This responsibility does not end with the completion of fieldwork or a research contract but is extended by the traces their work and presence leave behind in resource arenas – including personal and professional relations and obligations, and their actual research outputs. To that end, Shannon Morreira (2012) argues that consideration of the ethics of anthropological engagement in the field should extend to the ethics of dis-engagement after research activities are completed.

Most recent debates about positionality in the anthropology of resource extraction move away from a focus on positionality as "a matter of individual ethics and engagements" (Le Meur 2015: 406), and towards the broader implications of working in "spaces created by conflict" (Coumans 2011). In his text on the anthropological ethics of "mining encounters" (Pijpers and Eriksen 2018), Golub (2018) suggests that when assessing the ethics of their own positionalities in resource arenas anthropologists would do well to reflect on their entanglements and responsibilities along the four parameters proposed by Iris Young (2006: 127) in her work on structural positioning: *power, privilege, interest,* and *collective ability.* Several of his points warrant mention here. Whilst there remains a widespread scepticism about anthropology's capacity to change unjust social processes associated with resource extraction, the discipline's record shows that, in certain cases, research engagements do provide opportunities to affect outcomes, shape public opinion, or influence the actions of one or more groups occupying resource arenas (e.g. Kirsch 2018; Sawyer 2004) – and anthropologists must recognize the kind of power that comes with their engagements both in and with the field of resource extraction. Anthropologists ought to also reflect on their own privilege, and the structural factors that create and maintain it – which typically places them in a dominant position vis-à-vis their interlocutors. While "studying up" can challenge this dynamic (Sedgwick 2017), for those anthropologists whose careers are forged in resource encounters, "recognition of [their] privilege should affect what sort of work [they] do" (Golub 2018: 31). Finally, Golub points out that most anthropologists do not have a direct stake in the outcome of extractive operations and will not experience the positive or negative effects brought about by their work. Immersion in the field is always incomplete and only ever temporary – even for devotees of immersive ethnography. To that end, he suggests that "[r]ecognizing a *lack* of anthropological interest in mining encounters is key to respecting the autonomy and responsibility for local stakeholders to make their own decisions", and further, that "an analysis of the parameters of interest may ... reveal that simply because an anthropologist has the power to intervene in a mining encounter does not mean that they should do so" (ibid.: 32). The question remains whether it is possible for anthropologists to work in those contexts without "intervening" – and if so – whether, considering the relative power and privilege enjoyed by anthropologists, there can be a moral and ethical argument for a "disinterested anthropology" of resource extraction.

6 Conclusions

In an increasingly interconnected and "overheated" world (Eriksen 2016), anthropologists must come to grips with the multiple flows and dislocations that characterize resource arenas as multiplex and multisite fields of research, in order to provide sophisticated accounts of extraction. And while many anthropologists are cautious about their power and efficacy in these contexts, they are often better placed than most to build research alliances and access diverse sites, make sense of the different social dimensions of extraction, and identify and challenge unjust social processes.

In this chapter we have considered some of the positions occupied by anthropologists in resource arenas, and placed them in the wider context of the changing field of resource extraction. We considered the implications that different positions create for methodology and knowledge, and outlined ways in which anthropologists have debated the ethical dilemmas posed by working in these contested spaces. The diversity of anthropological literature presented here, and the positional stances assumed by anthropologists, points to two things. Firstly, they highlight the complex and dynamic nature of resource extraction and the versatility of anthropological expertise in extractive contexts. Secondly, they demonstrate the multifaceted moral and ethical implications associated with working in resource arenas. This means that, while the field evolves, new methodologies and forms of engagement emerge, and the next generation of anthropologists are drawn to resource arenas, Ballard and Banks' (2003: 306) call for "sustained reflection on the implications and consequences of our interventions" will continue to reverberate through the discipline.

Bibliography

Abu-Lughod L. (1990) Can there be a feminist ethnography? *Women & Performance: A Journal of Feminist Theory* 5 (1): 7–27.

Aiello L. (2012) Letter from the President of the Wenner-Gren Foundation. *Current Anthropology* 53 (S5): v.

Asad T. (1986) The concept of cultural translation in British social anthropology. In: Clifford J. and Marcus G.G. (eds) *Writing culture: The poetics and politics of ethnography.* Berkeley: University of California Press, 141–164.

Bainton N. (2010) *Lihir destiny: Cultural responses to mining in Melanesia*, Canberra: ANU Press.

Bainton N. (2020) Mining and Indigenous peoples, *Oxford Research Encyclopedia of Anthropology*, Oxford: Oxford University Press.

Bainton N.A., Ballard C., Gillespie K. and Hall N. (2011) Stepping stones across the Lihir Islands: Developing cultural heritage management in the context of a gold-mining operation. *International Journal of Cultural Property* 18 (1): 81–110.

Bainton N. and Owen J.R. (2019) Zones of entanglement: Researching mining arenas in Melanesia and beyond. *The Extractive Industries and Society* 6 (3): 767–774.

Bainton N., Owen J.R., Kenema S. and Burton J. (2020) Land, labour and capital: Small and large-scale miners in Papua New Guinea. *Resources Policy* 68: 101805.

Bainton N. and Skrzypek E. (2021) *Absent presence: Resource extraction and the state in Papua New Guinea and Australia*, Canberra: ANU Press.

Ballard C. and Banks G. (2003) Resource wars: The anthropology of mining. *Annual Review of Anthropology* 32: 297–313.

Banks G. (2013) Little by little, inch by inch: Project expansion assessments in the Papua New Guinea mining industry. *Resources Policy* 38 (4): 688–695.

Bell J. (2014) The veracity of form: Transforming knowledges and their forms in the Purari Delta of Papua New Guinea. In: Silverman R. (ed.) *Museum as process: Translating local and global knowledges.* London: Routledge, 105–122.

Bensen P. and Kirsch S. (2010) Capitalism and the politics of resignation. *Current Anthropology* 51 (4): 459–486.

Burton J. (1999) Evidence of new competencies? In: Filer C. (ed). *The social and economic impact of the Porgera gold mine.* Canberra: ANU Press.

Burton J. (2007) The anthropology of personal identity: Intellectual property rights issues in Papua New Guinea, West Papua and Australia. *TAJA* 18 (1): 40–55.

Burton J. (2014) Agency and the "Avatar" narrative at the Porgera gold mine, Papua New Guinea. *Journal de la Société des Océanistes* 138–139: 37–52.

Carstens P. (2001) *In the company of diamonds: De Beers, Kleinzee, and the control of a town*, Athens: Ohio University Press.

Cefkin M. (2012) Close encounters: Anthropologists in the corporate arena. *Journal of Business Anthropology* 1 (1): 91–117.

Clifford J. and Marcus G.E. (eds) (1986) *Writing culture: The poetics and politics of ethnography*, Berkeley: University of California Press.

Cochrane G. (2017) *Anthropology in the mining industry: Community relations after Bougainville's civil war*, Cham: Palgrave Macmillan.

Coleman S. and Collins P. (2006) *Locating the field: Space, place and context in anthropology*, Oxford: Berg.

Coumans C. (2011) Occupying spaces created by conflict: Anthropologists, development NGOs, responsible investment, and mining. *Current Anthropology* 52 (S3): 29–43.

Coumans C. (2017) Do no harm? Mining industry responses to the responsibility to respect human rights. *Canadian Journal of Development Studies / Revue canadienne d'études du développement* 38 (2): 272–290.

Coumans C. (2018) Into the deep: Science, politics and law in conflicts over marine dumping of mine waste. *International Social Science Journal* 68: 303–323.

Cunliffe A.L. and Alcadipani R. (2016) The politics of access in fieldwork: Immersion, backstage dramas, and deception. *Organizational Research Methods* 19 (4): 535–561.

D'Angelo L. and Pijpers R.J. (2018) Mining temporalities: An overview. *The Extractive Industries and Society* 5 (2): 215–222.

Dalsgaard S. (2019) "Seeing" Papua New Guinea. *Social Analysis* 63 (1): 44–63.

Dashwood H. (2012) *The rise of global corporate social responsibility: Mining and the spread of global norms*, Cambridge: Cambridge University Press.

De la Cadena M. (2015) *Earth beings: Ecologies of practice across Andean worlds*, Durham: Duke University Press.

De Rijke K. (2013) Coal seam gas and social impact assessment: An anthropological contribution to current debates and practices. *Journal of Economic & Social Policy* 15 (3): 29–58.

Dolan C. and Rajak D. (2016) *The anthropology of corporate social responsibility*, London and New York: Berghahn Books.

Douglas M. (1966) *Purity and danger: An analysis of concepts of pollution and taboo*, London and New York: Routledge & Keagan Paul.

Dwyer P.D. and Minnegal M. (2014) Where all the rivers flow west: Maps, abstraction and change in the Papua New Guinea lowlands. *The Australian Journal of Anthropology* 25 (1): 37–53.

Epstein A.L. (1958) *Politics in an urban African community*, Manchester: Manchester University Press.

Eriksen T.H. (2016) *Overheating: Coming to terms with accelerated change*, London: Pluto Press.

Eriksen T.H. (2018) *Boomtown: Runaway globalisation on the Queensland coast*, London: Pluto Press.

Espig M. and de Rijke K. (2018) Energy, anthropology and ethnography: On the challenges of studying unconventional gas developments in Australia. *Energy Research & Social Science* 45: 214–223.

Ferguson J. (1999) *Expectations of modernity: Myth and meanings of urban life on the Zambian Copperbelt*, Berkeley: University of California Press.

Ferry E.E. and Limbert M.E. (2008) *Timely assets: The politics of resources and their temporalities*, Santa Fe: School for Advanced Research Press.

Filer C. (1996) Global alliances and local mediations. *Anthropology Today* 12 (5): 26–27.

Filer C. (1999) The dialectics of negation and negotiation in the anthropology of mineral resource development in Papua New Guinea. In: Cheater A. (ed.) *The anthropology of power: Empowerment and disempowerment in changing structures*. London: Routledge, 88–102.

Filer C. (2018) Extractive industries and development. In: Callan H. (ed.) *The international encyclopedia of anthropology*. Wiley online.

Foster R.J. (2017) The corporation in anthropology. In: Spicer A. and Baars G. (eds) *The corporation: A critical, multi-disciplinary handbook*. Cambridge: Cambridge University Press, 111–133.

Geer S. (2003) The tricky/trickster role of the anthropologist: Ethical dilemmas of the consultant anthropologist in Papua New Guinea. *The University of Western Ontario Journal of Anthropology* 11 (1): 40–50.

Gilberthorpe E. (2013) Community development in Ok Tedi, Papua New Guinea: The role of anthropology in the extractive industries. *Community Development Journal* 48 (3): 466–483.

Girardet H. (2001) Darell Posey: Anthropologist who championed the rights of Amazonian tribes. *The Guardian*, 30 March 2001. Accessed online in December 2020 at: https://www.theguardian.com/news/2001/mar/30/guardianobituaries.

Gluckman M. (1961) Anthropological problems arising from the African industrial revolution. In: Southall A. (ed.) *Social change in modern Africa*. Oxford: Oxford University Press, 67–82.

Godoy R. (1985) Mining: Anthropological perspectives. *Annual Review of Anthropology* 14: 199–217.

Goldman L. (2000) *Social impact analysis: An applied anthropology manual*, Oxford: Berg.

Golub A. (2014) *Leviathans at the gold mine: Creating Indigenous and corporate actors in Papua New Guinea*, Durham: Duke University Press.

Golub A. (2018) From allegiance to connection: Structural injustice, scholarly norms and the anthropological ethics of mining encounters. In: Pijpers R.J. and Eriksen T.H. (eds) *Mining encounters: Extractive industries in an overheated world*. London: Pluto Press, 21–37.

Golub A. and Rhee M. (2013) Traction: The role of executives in localising global mining and petroleum industries in Papua New Guinea. *Paideuma* 59: 215–235.

Goodale M. (2006) Introduction to "Anthropology and human rights in a new key". *American Anthropologist* 108 (1): 1–8.

Gordon R.J. (1977) *Mines, masters and migrants: Life in a Namibian compound*. Johannesburg: Ravan Press.

Gupta A. and Ferguson J. (1997) Discipline and practice: "The field" as site, method, and location in anthropology. In: Gupta A. and Ferguson J. (eds) *Anthropological locations: Boundaries and grounds of a field science*. Berkeley: University of California Press, 1–46.

Hannerz U. (2010) Field worries: Studying down, up, sideways, through, backward, forward, early or later, away and at home. In: Hannerz U. (ed.) *Anthropology's world: Life in a twenty-first-century discipline*. London and New York: Pluto Press, 59–86.

Harris M. (1959) Labour emigration among the Mozambique Thonga: Cultural and political factors. *Africa* 29: 50–65.

High M.M. (2017) *Fear and fortune: Spirit worlds and emerging economies in the Mongolian gold rush*. Ithaca, NY: Cornell University Press.

High M.M. and Smith J.M. (2019) Introduction: The ethical constitution of energy dilemmas. *Journal of the Royal Anthropological Institute* 25 (S1): 9–28.

Hyndman D. (1995) *Ancestral rainforests and the mountain of gold: Indigenous peoples and mining in New Guinea*, Boulder, CO: Westview Press.

Idemudia U. (2010) Corporate social responsibility and the rentier Nigerian state: Rethinking the role of government and the possibility of corporate social responsibility in the Niger Delta. *Canadian Journal of Development Studies* 30: 131–153.

Jacka J. (2015) *Alchemy in the rain forest: Politics, ecology, and resilience in a New Guinea mining area*, Durham: Duke University Press.

Jacka J.K. (2018) The anthropology of mining: The social and environmental impacts of resource extraction in the mineral age. *Annual Review of Anthropology* 47 (1): 61–77.

Jackson M. (2013) *Lifeworlds: Essays in existential anthropology*. Chicago: University of Chicago Press.

Jackson R. (1991) Not without influence: Villages, mining companies, and government in Papua New Guinea. In: Connell J. and Howitt R. (eds) *Mining and Indigenous peoples in Australasia*. Sydney: Sydney University Press, 18–34.

Jalbert K., Willow A., Casagrande D. and Paladino S. (eds) (2017). *ExtrACTION: Impacts, engagements, and alternative futures*. London: Routledge.

Kapelus P. (2002) Mining, corporate social responsibility and the "community": The case of Rio Tinto, Richard's Bay minerals and the Mbonambi. *Journal of Business Ethics* 39: 275–296.

Kemp D. and Owen J.R. (2018) The industrial ethic, corporate refusal and the demise of the social function in mining. *Sustainable Development* 26 (5): 491–500.

Kirsch S. (1996) Anthropologists and global alliances. *Anthropology Today* 12 (4): 14–16.

Kirsch S. (2002) Anthropology and advocacy: A case study of the campaign against the Ok Tedi mine. *Critique of Anthropology* 22 (2): 175–200.

Kirsch S. (2014) *Mining capitalism: The relationship between corporations and their critics*, Oakland: University of California Press.

Kirsch S. (2018) *Engaged anthropology: Politics beyond the text*, Oakland: University of California Press.

Kirsch T.G. (2007) An appeal for principled symmetry: Anthropologies in South Africa and elsewhere. *Anthropology Southern Africa* 30 (3–4): 97–100.

Le Meur P.Y. (2015) Anthropology and the mining arena in New Caledonia: Issues and positionalities. *Anthropological Forum* 25 (4): 405–427.

Lee C.K. (2009) Raw encounters: Chinese managers, African workers and the politics of casualization in Africa's Chinese enclaves. *The China Quarterly* 199: 647–666.

Letté, M. (2009) Débordements industriels dans la cité et histoire de leurs conflits aux XIXe et XXe siècles. *Documents pour l'histoire des techniques* 17: 163–173.

Li F. (2015) *Unearthing conflict: Corporate mining, activism, and expertise in Peru*, Durham: Duke University Press.

Love T. and Isenhour C. 2016. Energy and economy: Recognizing high-energy modernity as a historical period. *Economic Anthropology* 3 (1): 6–16.

Luning S. (2014) The future of artisanal miners from a large-scale perspective: From valued pathfinders to disposable illegals? *Futures* 62: 67–74.

Macdonald G. (2002) Ethnography, advocacy and feminism: A volatile mix. A view from a reading of Diane Bell's *Ngarrandjeri Wurruwarrin. The Australian Journal of Anthropology* 13 (1): 88–110.

Magubane B. (1971) A critical look at indices used in the study of social change in colonial Africa. *Current Anthropology* 12: 419–445.

Malinowski B. (1961 [1922]) *Argonauts of the Western Pacific: An account of native enterprise and adventure in the archipelago of Melanesian New Guinea*. London: Routledge & Keegan Paul.

Marcus G.E. (1995) Ethnography in/of the world system: The emergence of multi-sited ethnography. *Annual Review of Anthropology* 24 (1): 95–117.

Marcus G.E. (1997) The uses of complicity in the changing mise-en-scene of anthropological fieldwork. *Representations* 59: 85–108.

Minnegal M. and Dwyer P.D. (2017) *Navigating the future: An ethnography of change in Papua New Guinea*, Acton, Australia: Australian National University Press.

Mitchell C.J. (1959) *The kalela dance: Aspects of social relationships among urban Africans in Northern Rhodesia*, Rhodes-Livingstone Papers, Manchester: Manchester University Press.

Morreira S. (2012) Anthropological futures? Thoughts on social research and the ethics of engagement. *Anthropology Southern Africa* 35 (3–4): 100–104.

Nader L. (1974) Up the anthropologists – perspectives gained from studying up. In: Hymes D. (ed.) *Reinventing anthropology*. New York: Pantheon Books, 284–311.

Nader L. (1997) Sidney W. Mintz lecture for 1995: Controlling processes tracing the dynamic components of power. *Current Anthropology* 38 (5): 711–738.

Nash J. (1979) *We eat the mines and the mines eat us: Dependency and exploitation in Bolivian tin mines*, New York: Columbia University Press.

Nugent S. and Shore C. (2002) *Elite cultures: Anthropological perspectives*, London: Routledge.

Ortner S.B. (1995) Resistance and the problem of ethnographic refusal. *Comparative Studies in Society and History* 37 (1): 173–193.

Ortner S.B. (2010) Access: Reflections on studying up in Hollywood. *Ethnography* 11 (2): 211–233.

Owen J. and Kemp D. (2017) *Extractive relations: Counterveiling power and the global mining industry*, London: Routledge.

Pijpers R.J. and Eriksen T.H. (2018) *Mining encounters: Extractive industries in an overheated world*, London: Pluto Press.

Rajak D. (2011) *In good company: An anatomy of corporate social responsibility*, Palo Alto, CA: Stanford University Press.

Richardson T. and Weszkalnys G. (2014) Introduction: Resource materialities. *Anthropological Quarterly* 87 (1): 5–30.

Rolston J.S. (2014) *Mining coal and undermining gender: Rhythms of work and family in the American West*, New Brunswick, NJ: Rutgers University Press.

Rosen L. (1977) The anthropologist as expert witness. *American Anthropologist* 79 (3): 555–578.

Sawyer S. (2004) *Crude chronicles: Indigenous politics, multinational oil, and neoliberalism in Ecuador*, Durham: Duke University Press.

Sawyer S. and Gomez E.T. (2012) *Politics of resource extraction: Indigenous peoples, multinational corporations and the state*, London: Palgrave Macmillan.

Sedgwick M.W. (2017) Complicit positioning: Anthropological knowledge and problems of "studying up" for ethnographer-employees of corporations. *Journal of Business Anthropology* 6 (1): 58.

Skrzypek E. (2020) Extractive entanglements and regimes of accountability at an undeveloped mining project. *Resources Policy* 69: 101815.

Smith J.M. (2019) Boom to bust, ashes to (coal) dust: The contested ethics of energy exchanges in a declining US coal market. *Journal of the Royal Anthropological Institute* 25 (S1): 91–107.

Smith J. and High M.M. (2017) Exploring the anthropology of energy: Ethnography, energy and ethics. *Energy Research & Social Science* 30: 1–6.

Stewart P. and Strathern A. (2005) *Anthropology and consultancy: Issues and debates*, New York: Berghahn Books.

Strathern, A. and Stewart, P.J. (2001) Introduction: Anthropology and consultancy: Ethnographic dilemmas and opportunities. *Social Analysis* 45 (2): 3–22.

Straube C. (2020) Speak, friend, and enter? Fieldwork access and anthropological knowledge production on the Copperbelt. *Journal of Southern African Studies* 46 (3): 399–415.

Stryker R. and González R.J. (2014) *Up, Down, and Sideways: Anthropologists Trace the Pathways of Power*, New York: Berghahn Books.

Szeman I.E. (2017) Introduction: Pipeline politics. *South Atlantic Quarterly* 116 (2): 402–407.

Taussig M. (1980) *The devil and commodity fetishism in South America*, Chapel Hill: The University of North Carolina Press.

Trigger D. (2011) Anthropology pure and profane: The politics of applied research in Aboriginal Australia. *Anthropological Forum* 21 (3): 233–255.

Tsing A. (2005) *Friction: An ethnography of global connection*, Princeton, NJ: Princeton University Press.

Voyles T.B. (2015) *Wastelanding: Legacies of uranium mining in Navajo Country*, Minneapolis: University of Minnesota Press.

Webb-Gannon C. (2014) Merdeka in West Papua: Peace, justice and political independence. *Anthropologica* 56 (2): 353–367.

Weiner J.F. (2002) Assuming the mercenary position: Changing perspectives in long-term fieldwork. *The Asia Pacific Journal of Anthropology* 3 (2): 33–43.

Welker M.A. (2014) *Enacting the corporation: An American mining firm in post-authoritarian Indonesia*, Berkeley: University of California Press.

Welker M. and Wood D. (2011) Shareholder activism and alienation. *Current Anthropology* 52 (S3): S57–S69.

Weszkalnys G. (2010) Re-conceiving the resource curse and the role of anthropology. *Suomen Antropologi, Journal of the Finnish Anthropological Association* 35 (1): 87–90.

Wilson E. and Swiderska K. (2009) *Extractive industries and Indigenous peoples in Russia: Regulation, participation and the role of anthropologists*, London: International Institute for Environment and Development.

Young I.M. (2006) Responsibility and global justice: A social connection model. *Social Philosophy and Policy* 23 (1): 102–130.

Zhouri A. (2017) Introduction: Anthropology and knowledge production in a "minefield". *Vibrant: Virtual Brazilian Anthropology* 14 (2): 1–9.

9

SUSTAINABILITY

Cristiano Lanzano

1 Introduction

The extractive sector is classically associated with pollution, landscape change, exhaustion of non-renewable resources, and structural transformations, possibly leading to economic dependency or unfair distribution of wealth. This, it seems, is the opposite of scholarly definitions of environmental and social sustainability, and of what emerges in the political field when institutional actors and social movements negotiate their views of "sustainable development". For these reasons, the expression "sustainable mining" can sound like a contradiction in terms, and it has often been considered by both activists (for example Whitmore 2006) and anthropologists (Kirsch 2010) as an oxymoron designed by corporations to conceal the harm they produce and neutralize critics.

Nevertheless, as will transpire in this chapter, the mining sector has progressively multiplied efforts to legitimize itself and to engage with sustainability debates, leading to the creation of hybrid fields where collective development goals and private interests are framed in similar terms. From assessment and transparency procedures imposed on companies, through voluntary standards and programmes, to civil society initiatives aimed at making the mining sector fairer, anthropologists have observed the manifestations of such a paradox and analysed the contradictory political consequences it has generated.

This chapter reviews anthropological analysis of the mining/sustainability nexus, predominantly published during the last two decades. In Section 2, I begin by briefly describing the initial emergence of the sustainability agenda in the extractive industries at the turn of the century. Sections 3 and 4 review the anthropological literature on forms taken by the sustainability agenda in the mining sector, namely, environmental impact assessment, corporate social responsibility and "fair" mining. In Section 5, I refer to the anthropological scholarship on temporality and future-

DOI: 10.4324/9781003018018-9

making, which offers an alternative way to approach questions of sustainability in connection to resource extraction.

2 The emergence of the sustainability agenda in the extractive sector

The notions of "sustainability" and "sustainable development" first appeared in global politics in the 1980s.[1] Their outreach grew constantly over the following decades, nourished by global summits, international treaties, and a flourishing scholarly literature that highlighted the environmental and social consequences of capitalistic growth and dominant models of development, and that strove to imagine and theorize solutions or alternative futures.[2] The relevance of sustainability for the extractive sector became more explicit at the end of the 1990s, after a decade when debates about globalization and the "new world order" had led to increased public attention being paid to the power and impacts of transnational corporations, particularly in the extractive sector. Cases like the protests and lawsuits over the BHP Ok Tedi mine in Papua New Guinea (Banks and Ballard 1997; Kirsch 2014) and the execution of activists campaigning against Shell operations in Nigeria (Watts 2005) contributed to the deteriorating reputation of oil and mining companies. The question of conflict minerals – the role of mineral extraction and trade in financing civil wars and armed groups – received wide media coverage and stimulated campaigning by civil society organizations.

As a response, companies and international organizations launched initiatives aimed at linking the extractive sector to the sustainability agenda. The World Bank (WB) group initiated a review process to assess its own investments in the extractive sector, and published the report *Striking a Better Balance* (WB 2004) in which it mentioned the relevance of poverty reduction and sustainable development, and designed a set of conditions requiring transparency and monitoring, stakeholder involvement, and mechanisms for mitigating social and environmental risks. At about the same time, a group of mining company CEOs constituted the GMI (Global Mining Initiative) and launched the Mining, Minerals and Sustainable Development (MMSD) project. MMSD published the report *Breaking New Ground* (MMSD 2002), considered a milestone for the engagement of corporate mining actors with debates about development and sustainability (see Owen and Kemp 2013). In 2002, the Plan of Implementation adopted by the World Summit on Sustainable Development in Johannesburg included an article on the contribution of the mining industry to sustainable development, showing the success of the initiatives of the previous years. The text called stakeholders to "support efforts to address the environmental, economic, health and social impacts and benefits of mining", to "enhance the participation of stakeholders, including local and indigenous communities and women" and to "foster sustainable mining practices" (UN 2002: 37).

At a general level, criticism of the vague and contradictory character of the concepts of "sustainability" and "sustainable development" appeared early (Lélé

1991). Debates took place between those who prioritized a narrower definition of ecological sustainability and those who lobbied for the consideration of social and political dimensions. The Brundtland report, published in 1987, was noted for its peculiar combination of radical and reformist approaches, highlighting the intertwining of ecological sustainability and global development, but also calling for greater economic growth (Robinson 2004). These controversies disillusioned many observers and raised concern over the possible co-optation of the notion.

Meanwhile, in the extractive sector, activists and indigenous organizations also criticized institutional and corporate narratives of mining sustainability (Power 2002; Whitmore 2006). Nonetheless, these began to materialize in a new set of practices – both compulsory and voluntary – showcased by mining companies to demonstrate their commitment to social and environmental issues. These included more or less formalized negotiation procedures with local communities, impact assessment, and disclosure or "sustainability accounting" that were, however, variously interpreted and implemented (Jenkins and Yakovleva 2006), thereby demonstrating the limits of the "audit culture" associated with corporate accountability (Kemp et al. 2012; Lodhia and Hess 2014).

3 Assessing sustainability

Environmental impact assessment (EIA) procedures, although introduced by national legislation in some countries as early as the 1970s, did not become standard across the world until the 1990s, when international organizations, and ultimately the Rio declaration in 1992, recommended their adoption (Morrison-Saunders and Arts 2004). By the turn of the century, EIAs had become mandatory in most countries, including those in the Global South, then experiencing a concentration of new mining investment. EIAs became linked with the growing reflection on sustainability and progressively absorbed some of its elements, integrating the consideration of social impacts, for example, and adopting participatory processes (Joyce and MacFarlane 2001).

In the fields created by EIAs and other forms of impact assessment, different visions of sustainability and human–environmental relations interact, and conflicts between investors, political authorities, and local communities are played out. Anthropologists have observed that EIA processes can become detached from their theoretical mandate, failing to acknowledge environmental risks or reducing the participation of stakeholders to merely "cosmetic" actions. In line with an established critique of "participation" in development studies (see Cooke and Kothari 2001; Rahnema 1992), Perrault (2015) describes how public hearings and consultations undertaken after protests against mining pollution in the region of Oruro (Bolivia) had an essentially performative function, limiting spaces for dissent and reconfirming the authorities' commitment to extractive projects.

Given that large mining projects often affect the livelihood of indigenous groups and rural communities, EIAs take place in contexts characterized by highly unequal relations, in material as well as ontological terms. To the extent that their frameworks rest on modernist assumptions and normative conceptions of economic development, impact assessments may reinforce those inequalities, marginalizing ways of living, or silencing ways of knowing – and talking about – the environment. Westman (2013) describes how development-driven visions of the future have biased EIAs conducted before tar (oil) sands mining projects and overlooked the foraging economies and the associated worldviews of indigenous groups living in northern Alberta (Canada). The misrepresentation of reindeer herding practices in the EIA carried out for the opening of a copper mine in northern Sweden also instigated a "knowledge controversy" that brought Sami communities and the mining company into opposition (Lawrence and Kløcker Larsen 2017). Even when EIAs are not conducted by companies to obtain clearance for their large-scale operations, but instead required of small-scale miners in order to obtain licenses, the hegemony established by the technical expertise and language in which they are articulated discriminates against more vulnerable producers, as Spiegel (2017) shows in the case of the Zimbabwean artisanal mining sector.

Anthropologists have also analysed how the approaches employed by the authors of EIAs shape their understanding of risks and impacts, and the practical consequences of these framings. As argued by Teixeira et al. (2021) in regard to mining and hydroelectric projects in the Brazilian state of Minas Gerais, the priority given to quantifiable data is aimed at rendering territories easily manageable and producing a specific "economy of visibilities" that favours the implementation of industrial development and the suppression of conflict. As Li (2011) shows in the case of the Pascua Lama gold mine in Chile, standardized techniques for calculating environmental risks and impacts and designing corresponding compensation policies serve the "logic of equivalence that makes the possible consequences of a mining project commensurate with the mining company's mitigation plans" (Li 2011: 62).

In line with Ferguson's work on development (Ferguson 1990), many anthropologists have highlighted how impact assessments tend to de-politicize resource-related conflicts and frame them as merely technical problems. This is also reflected in the broader tendency of political authorities to withdraw and delegate responsibility for socio-environmental consequences to investors and companies. These, in turn, resort increasingly to consultant firms that can act as facilitators, and absorb the risks associated with negotiations with local communities, conducting de facto political mediation that exceeds their technical mandate (Dougherty 2019; Li 2009). However, this technicization is often resisted by affected communities, who can appropriate assessment procedures in order to counter pro-mining narratives (Lawrence and Kløcker Larsen 2017; Li 2011). Developing Foucault's understanding of how apparatuses of power also enable – while constraining – possibilities of resistance, Aguilar-Støen and Hirsch (2017) describe how protestors against the gold mining project El Tambor in Guatemala built a successful coalition of local NGOs and international experts that challenged the favourable EIA on technical and judicial grounds. In other terms, impact assessments can function as technologies for the domestication of risks and the suppression of conflicts, but they are occasionally turned into sites of struggle (Nielsen 2017) where discourses of mining sustainability are re-politicized.

4 Promoting sustainable development? Anthropological engagements with CSR and "fair" mining

In recent decades, corporate social responsibility (CSR) has become an object of analysis for anthropologists, particularly those interested in the development and adaptation of global capitalism and its localized manifestations (Dolan and Rajak 2016). The emergence and consolidation of CSR protocols in the mining industry since the 1990s (see also Shever this volume) overlapped with the debate on sustainability and on the environmental impacts of mining. Embodying corporations' displayed commitment to the social components of sustainability and to the promotion of sustainable development, CSR discourses articulate with notions of accountability and responsibility that reflect the changing moral regimes of capitalist production (Dolan and Rajak 2011). Standardized CSR protocols may lead to the adoption of technocratic and supposedly neutral approaches to issues, such as land compensation, environmental damage, social inequality, and distribution of benefits, that are eminently political (Benson and Kirsch 2010; Dolan and Rajak 2016; Sharp 2006). In this sense, critiques of CSR converge with those concerning the depoliticizing effects of dominant visions of development (Ferguson 1990; Sydow 2016) and sustainability.

Furthermore, the translation of CSR and sustainability protocols into regulations and standardized frameworks can lay the groundwork for companies' disengagement from the social and environmental consequences of their activities. Studying an India-based diamond manufacturing facility working as a subcontractor of De Beers, Cross (2011) argues that voluntary regulations, ethical and environmental standards, and audit procedures created by a company to promote best practices in its supply chain end up, paradoxically, as tools for the reinforcement of an "ethic of detachment" that allows it to limit its responsibility and its relations with producers. Similarly, as argued by Gardner et al. (2012), community partnerships and employment policies supposed to be responsive to local needs actually allowed Chevron to disengage from the claims and expectations of people living in the surroundings of its gas extraction operations in Bibiyana (Bangladesh). Sustainable development programmes sponsored by companies thus act as technologies to deflect attention from conflicts and defuse responsibility, as argued by Warnaars (2012) of the awareness campaigns developed by Corriente in the copper-producing site of El Pangui (Ecuador); or to produce a shift from the principle of corporate accountability to a more consensual notion of "shared responsibility", as described by Li (2015) of health-related training implemented by Doe Run in the heavily polluted smelter town of La Oroya (Peru).

As with any programme in the development sector, the different interventions designed by companies to reassert their commitment to promote environmentally and socially sustainable forms of mining need to identify their beneficiaries.

Departments of company–community relations, the delimitation of affected villages and territories, and guidelines for stakeholders' participation, all contribute to the negotiation of social, territorial, and demographic boundaries that define obligations and entitlements variously connected with pre-existing notions of community or belonging (see Pijpers this volume). Defining significant communities can expose mining companies to criticism and reveal contrasting views of their obligations, as exemplified in the strategies of community development adopted by Rio Tinto in South Africa (Kapelus 2002). In many cases, mining companies' policies have encouraged a tightening of boundaries between social units, as Gilberthorpe (2007) observes of the Kutubu oil extraction project in Papua New Guinea, where porous and inclusive solidarity networks were progressively replaced by stricter patrilineal criteria for group membership as a consequence of compensation policies implemented by the company. They have also led to restrictions on mobility; for instance, AngloGold Ashanti's attempts to define a traditional mining community in the vicinities of its Guinean mines reinforced an exclusionary logic of autochthony and the marginalization of itinerant gold miners previously integrated into the local economy (Bolay 2014).

Beyond CSR schemes and corporate-led programmes, other attempts have been made to promote sustainable practices in the mining sector, bypassing corporations or focusing on contexts where their presence is marginal, such as informal or small-scale mining: namely, initiatives inspired by the principles of ethical consumption and fair trade (Fisher 2018a). From the 2000s, awareness campaigns on conflict minerals – initially focused on diamonds and later extended to other minerals such as gold or the so-called 3Ts (tantalum, tungsten, and tin) – created space for NGOs and retailers to work towards mineral value chains that could be certified as conflict-free, ethical, or sustainable (Hilson 2008; Van Bockstael 2018). More recently, initiatives in this direction have mainly targeted artisanal and small-scale gold mining, building sophisticated networks of local and international NGOs, and certifying entities, retailers' associations, and other actors in the fair trade sector. However, the application of the fair trade movement's general principles – originally elaborated in reference to handicrafts and agriculture – to the small-scale mining sector has required adaptation and is still contested (Childs 2014; Hilson 2008).

Fair trade is valued for its attempt to reconcile the needs and motives of Northern consumers with the livelihoods and rights of Southern producers: in other words, to uncover mechanisms of exploitation concealed by everyday practices of consumption, and to challenge, in Marxian terms, the routines of "commodity fetishism" (Fridell 2006; Goodman 2004; Luning 2013; Luning and de Theije 2014). Nevertheless, critical analyses have shown that notions of fairness and traceability are contested and negotiable (Childs 2008; Marston 2013). The way in which standards, charters, and certification schemes defined at an international level materialize to match the realities of artisanal miners on the ground is a crucial issue. Sarmiento et al. (2013) describe the case of the Oro Verde project in the Colombian region of Chocó as a positive experience resting on a structured alliance between different actors – community councils, NGOs, and research bodies – and on the resulting "institutional nesting". Other studies highlight problematic aspects. In East Africa, the initial phases of the Fairtrade and Fairmined project were accompanied by controversies over the fixing of prices for gold producers that undermined trust and questioned the contradictory interpretations of fairness (Fisher and Childs 2014). Entry requirements for projects can also limit the participation to formalized or well-networked producers, risking elite capture (Fisher 2018b; see also Hilson et al. 2016). In those cases, fair trade initiatives would appear to contradict their own pro-poor stance, not to mention their aesthetic idealization of smallness (Luning and de Theije 2014).

Beyond the long and complex histories that gave rise to them, CSR and fair trade represent two forms assumed by the mining/sustainability nexus. In the first, large corporations are supposed to minimize their negative impacts by applying voluntary measures. In the second, the power to produce change in the direction of sustainability is attributed to groups of organized consumers in alliance with retailers, producers' groups, and other organizations that are committed to the same principles of environmental and social justice. Developed during a time dominated by narratives of globalization and the supposed crisis of national states, these two fields have reinforced the role of private and non-governmental actors in the governance of extractive activities, the management of their social and environmental dimensions, even the provision of public goods in their surroundings. Resembling the local "arenas" described by development anthropologists (Bierschenk 1988; Olivier de Sardan 1993), these hybrid forms of governance function both through emerging standardized procedures that include negotiations and formal agreements (for example in Papua New Guinea or New Caledonia, see Filer and Le Meur 2017; Le Meur 2015), and through informal strategies of "forum shopping" and "shopping forum" that allow the mutual recognition and legitimization of corporate actors and customary authorities (for an example in Burkina Faso, see Luning 2010). However, the role of the state is not simply superseded by these developments. CSR programmes implemented by companies feed into pre-existing political alliances, clienteles, and dynamics of leadership that reconfigure relations between local and national authorities (on oil extraction in the Russian region of Perm, see Rogers 2012). They can also re-legitimize the state through narratives of "patriotic capitalism" and empowerment that elude issues of redistribution or social justice, as in the South African case described by Rajak (2011).

In recent years, beside CSR and fair trade, new sets of practices have emerged: due diligence principles requiring the traceability of value chains and intergovernmental initiatives promoting transparency (for the Extractive Industries Transparency Initiative, see Corrigan 2014; Smith et al. 2012); legislation imposing the certification of imported minerals (Cuvelier 2017; Cuvelier et al. 2014; Vogel and Raeymaekers 2016); ideas of "mining shared value" (Fraser 2019) and other ways of reframing the business case for sustainable development. In the future, these practices may partly replace previous attempts to push the sustainability agenda in the mining sector, but they will likely continue to raise questions on the voluntary or compulsory nature of regulations, the disconnection between general principles and their application on the ground, the moral regimes that they contribute to shape or to transform, and their effects on the broader political and social fields in which they become embedded.

5 Mining temporalities, sustainability, and (un)desirable futures

As reviewed so far, anthropologists have looked at political regulations, corporate strategies, and civil society efforts to translate the general principles of sustainability into practice, analysing procedures of impact assessment, CSR, and fair trade in the extractive industry. However, anthropology offers tools for looking above and beyond current mainstream interpretations of sustainability by embracing themes such as temporality, inter-generational justice, and the future in relation to resource consumption and extraction. Recent years have seen a surge in anthropological interest in time and future-making, and multiplying attempts to study people's ideas and projects for the future in connection with their capacity to aspire (Appadurai 2013). In delineating a research agenda for an anthropology of the future, Pels (2015) suggests identifying "futurisms", those typically modern conceptualizations of time that become manifest around a series of posited epochal ruptures, and replacing them with views facilitating multiple futures. Whether sustainability falls into his category of futurism, or can instead be resignified as a path toward multiple ontologies and plural temporalities (see also Brightman and Lewis 2017; Moore 2017), remains to be seen.

The explicit connection between future-making and a possible anthropology of sustainability was highlighted by Persoon and van Est, who noted that "the global sustainability debate, which explicitly addresses the future and even the long-term future" (Persoon and van Est 2000: 7), posed a challenge to a discipline like anthropology, traditionally more concerned with space and focused on the present. Ferry and Limbert (2008) have argued about the intrinsic temporal character of "resource imaginations", defined as the ideational systems resulting from specific ways of conceiving relations between human societies and their environment, and of turning specific objects or substances into resources. Natural resources do not only allow a reframing of histories and collective identities based on an idealized past, "[t]heir quality as potential wealth generates a possible future or futures, as well as the futures of those collectivities that lay claim to them or grapple with their limits and scarcity" (Ferry and Limbert 2008: 6).

In recent years, anthropologists of mining have contributed significantly to this elaboration. D'Angelo and Pijpers (2018) introduce the expression "mining temporalities" precisely in order to describe and reflect on the multiple entanglements of past, present, and future in the extractive industries, and on the political dimension of these same entanglements. As they point out:

> While it is true that many mining operations around the world can be interpreted as efforts to articulate or synchronize different temporalities (e.g.,

market cyclical temporalities and cycles of production), it is also true that there is always a nonsynchronizable temporal element (e.g., the long-term regenerative capacity of natural resources compared to the rapid mining rhythms of extraction).

(D'Angelo and Pijpers 2018: 215)

This has implications for how anthropology can contribute to a critical analysis of the mining/sustainability nexus. As theorists of "strong" sustainability[3] highlight, all societies must inevitably deal with the finitude of non-replaceable resources; however, they expand on different worldviews and practices which may negotiate with this environmental constraint and achieve, or aim for, modes of resource extraction that can be viewed and experienced as economically or socially sustainable (on "three-pillar" conceptions of sustainability, see Lozano 2008; Purvis et al. 2019). By looking at how different societies organize the pace of mining production and elaborating on imaginaries of resource consumption and future depletion, anthropologists can shed light on local and situated conceptualizations of the social and political-economic dimensions of mining (un)sustainability.

Mining economies often create expectations that can translate into shared modalities of imagining the future. Even before activities begin, the anticipation of resource extraction mobilizes both individual professionals and institutions for the purpose of assessing the geological and economic potential of mineral reserves, thereby generating forms of speculative knowledge (see Weszkalnys 2015, for the case of oil in São Tomé and Príncipe) and "futurework" (see Olofsson 2020, on mineral exploration in Sweden) aimed at dealing with the uncertainty and indeterminacy of future extraction and its profitability. These practices of anticipation have important social effects: they generate tensions and disagreements in the perception of landscapes and livelihoods, as in the case of an announced project for an open pit mine in Minnesota, US (Phadke 2018); they highlight differences in ideas of urban and mining development and in the associated temporalities between different social groups, illustrated by the prospects for coal mining in northern Mozambique (Wiegink 2018); and they coagulate both political support and opposition to mining through different politicized imaginaries and uses of time, as shown by the debates accompanying shale gas exploration in the English county of Lancashire (Szolucha 2018) and by popular resistance to coal mining in northwest Bangladesh (Chowdhury 2016). Contestation can arise even when the "forward-looking temporality" of mining is represented as a discontinuity and a step toward the energy transition, as in Portugal where plans for the extraction of lithium, seen as an essential component of a post-oil sustainable energy politics, highlight different views of ecologically and economically viable futures (Pusceddu 2020).

When promised progress fails to materialize, however, leaving people to deal with less optimistic futures, expectations are followed by disillusionment, as shown in Ferguson's seminal ethnography of life in the Zambian Copperbelt after the decline of copper production and the failure of promises of modernity and development (Ferguson 1999). Exposing the fallacy of imagined unilineal progress, post-boom phases of decline are themselves articulated and contradictory. Acceleration and deceleration can coexist in the same context and foster feelings of both hope and disenchantment, as demonstrated by the reopening and subsequent decrease in production of an iron ore mine in Sierra Leone (Pijpers 2016). Even during mining booms, the anticipation of resource depletion can generate anxiety or strategies for coping with future decline. In the Omani town studied by Limbert (2010), awareness of the exhaustibility of oil reserves resurfaced in both governmental narratives and the popular imagination, reinforcing a sense of uncertainty and fuelling the perception of contemporary prosperity as a dream-like interlude. Günel (2019) shows how, in Abu Dhabi, the perspective of a post-oil future has driven investment in an urban district dedicated to research and innovation in renewable and clean energy, nourishing imaginaries where technical adjustments would solve the problem of an unsustainable oil dependency without radically questioning the development model and the consumption levels associated with it.

Post-mining scenarios connect with issues of sustainability, not only through planning, anticipation, and adaptation to decline, but also, more obviously, when extractive activities effectively come to an end and the processes of mine closure pose the challenge of environmental remediation. Mining-related environmental disasters and highly polluting activities leave legacies of long-lasting health problems and social suffering, but also of environmental awareness and social justice activism, as ethnographic work on the aftermath of asbestos mining shows (for Italy, see Loher 2020; for Brazil, see Mazzeo 2021). Even in less dramatic cases, the ending of extractive activities leaves behind experiences of dispossession and epistemic injustice that hinder the participation of local or aboriginal communities in processes of remediation or post-mining planning (for the case of a gold mine closure in northern Canada, see Sandlos and Keeling 2016; for post-coal Appalachia, US, see Tarus et al. 2017). While the growing emphasis on sustainability has positively contributed to the institutionalization of participatory processes in post-mining remediation, Beckett and Keeling (2019) argue for a rethinking of the dominant technical approaches and suggest reframing mine closure as a question of environmental justice and care. Beside environmental aspects, anthropologists have also observed how the end of mining highlights contradictory visions of the future. Comparing the mining towns of Fria (Guinea Conakry) and Obuasi (Ghana), Knierzinger and Sopelle (2019) describe the effects of mine closures on diverging socio-political projects and the emergence of civil society groups advocating the advent of a new or a post-extractivist phase and claiming the right to the city. Analysing the imaginaries related to development and mining in Papua New Guinea, Halvaksz (2008) shows how, despite ongoing conflict, a desire for continued mining was common among Biangai communities even after the closure of the main industrial mine, which was experienced as an opportunity to negotiate new relations and identify new spaces of extraction.

The relevance of temporalities and future-making is also shown by studies on artisanal and small-scale mining economies. Here, too, mining can become an "economy of dreams" (Pijpers 2017), nourishing hopes of social mobility and perspectives of fortuitous earnings. Cycles in mineral production affect the ways in which informal miners experience time and plan their future in multiple ways. The alternation of boom and bust phases influences the gendered patterns of informal labour and exposes vulnerable categories to the corresponding, varying intensity of state control or property enforcement, as Mususa (2010) shows in her study of female informal traders in the new Zambian Copperbelt. Former boom towns can become "places of waiting" where people invest in alternative sources of livelihood while retaining the possibility of migrating and resuming their former activities, as observed by Walsh (2012) of a declining sapphire trade town in Madagascar. As I describe in the example of a gold mining site in western Burkina Faso, decline in production and demographic decrease can be turned into an opportunity by local miners and residents who strive to regain access to the resource, while consolidating informal but more durable governance and profit-sharing patterns that can survive short-lived booms (Lanzano 2018).

One of the peculiar aspects of artisanal mining is that the temporalities of global capitalism articulate more or less readily with different modes of production, social practices, and understandings of time. In analysing the work of artisanal miners in Sierra Leone, D'Angelo (2019) shows the contrast between human and non-human temporalities: on the one hand, miners attempt to articulate different extractive activities and other sources of livelihood (including farming), in order to produce a combination of spatio-temporalities securing immediate income while formulating projects for a more distant future; on the other, the non-human "deep time of resources" questions the sustainability of any extractive activity. Miners also experience dispossession and vulnerability in temporal terms, as in the case of 3T mines in eastern DRC studied by Smith (2011), where information asymmetry, volatility of prices, and, more generally, uncertain working conditions made the production of predictable time impossible.

Because of its partial continuity with pre-colonial or non-industrial extractive practices and the higher degree of its social embeddedness, small-scale mining offers material for thinking about time, sustainability, and finitude of resources in less normative or Eurocentric ways. Botchway (1995) argues that customary law and religious practices associated with pre-colonial gold mining in the forest regions of Ghana, while not being framed in conservationist terms, contributed to minimizing environmental degradation. While often represented as traditional and low-tech, artisanal mining economies are in reality characterized by change and innovation. Technological change is reflected in transformations of work and production rhythms, but also in the imagined temporal depth of mining and in miners' shared projection of their own activities onto the future, as shown in eastern Guinea Conakry by the shift from the system of *bè* – targeting gold alluvial deposits and managed customarily by village authorities – toward more intensive forms of hard-rock mining (Lanzano 2018; Lanzano and Arnaldi di Balme 2017). By making visible artisanal miners' voices, agency, and knowledge, anthropologists are well equipped to "understand better what futures miners anticipate and wish for, acknowledging the spatially and historical situated character of such anticipation" (Fisher et al. 2021: 191).

To be sure, not all the perceptions of time and elaborations of the future explored in the studies quoted above – whether involving artisanal miners, local communities, political authorities, or industry professionals – are explicitly linked to notions of social acceptability or environmental soundness. Yet, in its broadest sense, sustainability ultimately evokes desirable futures, inter-generational justice and the long-term viability of certain human endeavours and lifeways – whether maintained or forced to change (see Moore 2017). By researching the "multiple time frames" at work in mineral economies, and the practices of future-making in contexts marked by resource extraction, anthropologists can expand their reflection beyond the – necessary – critique of top-down political and corporate discourses, contributing to nuance the developmental "narratives of future positives" (Luning 2018: 283) and to de-centre notions of (mining) sustainability.

6 Conclusion

Reviewing a large number of documents and statements on sustainability within the mining industry, Han Onn and Woodley (2014) give an overview of the impressive variability in corporate uses of the concept. The polysemy and adaptability of the notion of sustainability have undeniably contributed to its success, making it easily appropriable by different extractive sector actors for different political and discursive uses. At the same time, its vagueness has attracted criticism, and its normative component – what kind of development is viable and desirable in the long run? – has led anthropologists to expose its contradictions.

It is perhaps not a coincidence that anthropologists who have investigated the mining/sustainability nexus have also been reflective about their positionality (see Bainton and Skrzypek this volume). Companies' attempts to secure a social license and to stabilize their relations with local communities rely on the work of anthropologists, with companies either employing them directly or contracting them for studies, thus producing various forms of institutional embeddedness (Coumans 2011). In a spectrum ranging from anti-mining advocacy to collaboration, anthropologists in this field must navigate complex situations often characterized by "fuzzier forms of embeddedness" (Welker 2014: 7). As Le Meur (2015) notes, the entry of discourses and practices of sustainability to the field of resource extraction has rendered mining arenas more similar to those previously studied by development anthropologists, including the problems connected to their positionality as consultants, applied scholars, or distanced critics. Anthropologists might not be able to decide once and for all whether "sustainable mining" is an attainable goal or an ideological smokescreen, but they can continue to reflect on the effects of this idea on their work and on its impact on the world at large.

Acknowledgements

I want to thank David Kideckel for precious reading suggestions, and Eleanor Fisher and Cecilia Navarra for reading earlier versions of this chapter. As editors of this volume, Lorenzo D'Angelo and Robert Pijpers have provided essential support and feedback.

Notes

1 The genealogies of the concepts of "sustainability" and "sustainable development" are as diverse and controversial as their meanings or political uses. See, for example, DuPisani 2006; Mebratu 1998; Robinson 2004.

2 Among the most important steps in the growth and mainstreaming of sustainability debates are mentioned: the 1972 publication of *The Limits to Growth* by the Club of Rome, which used computer modelling to elaborate on projections of trends in population growth and resource consumption, predicting serious crises by the end of the 21st century; the World Conservation Strategy, published in 1980 by the International Union for the Conservation of Nature (IUCN), where the term "sustainability" was first included; the Brundtland report (actually titled *Our Common Future*) published in 1987 by the World Commission on Environment and Development (WCED), which provided the first definition of "sustainable development" ("development that meets the needs of the present without compromising the ability of future generations to meet their own needs", WCED 1987: 43); the 1992 UN Conference on Environment and Development in Rio de Janeiro, which led to the production of many documents (such as the Rio Declaration or Agenda 21) where the concept of sustainable development was operationalized for the first time; and the summits marking the 10th and the 20th anniversaries of the Rio Conference (the 2002 World Summit on Sustainable Development in Johannesburg, and the 2012 UN Conference on Sustainable Development in Rio de Janeiro, which launched the Sustainable Development Goals). See Hopwood et al. 2005; Lélé 1991; Mebratu 1998; Robinson 2004.

3 "Strong" conceptualizations of sustainability postulate the non-substitutability of natural capital and the need to maintain its stock unchanged. Theorists of "weak" sustainability, on the other hand, see natural and manufactured capital as partially interchangeable and are open to the possibility of technology filling the gaps produced by humanity in the natural world. See Beckerman 1994; Hopwood et al. 2005.

Bibliography

Aguilar-Støen M. and Hirsch C. (2017) Bottom-up responses to environmental and social impact assessments: A case study from Guatemala. *Environmental Impact Assessment Review* 62: 225–232.

Appadurai A. (2013) *The future as cultural fact. Essays on the global condition*, London: Verso.

Banks G. and Ballard C. (1997) *The Ok Tedi settlement: Issues, outcomes and implications*, Canberra: ANU Press.

Beckerman W. (1994) Sustainable development: Is it a useful concept? *Environmental Values* 3 (3): 191–219.

Beckett C. and Keeling A. (2019) Rethinking remediation: Mine reclamation, environmental justice, and relations of care. *Local Environment* 24 (3): 216–230.

Benson P. and Kirsch S. (2010) Corporate oxymorons. *Dialectic Anthropology* 34: 45–48.

Bierschenk T. (1988) Development projects as arenas of negotiation for strategic groups: A case study from Bénin. *Sociologia Ruralis* 28 (2–3): 146–160.

Bolay M. (2014) When miners become "foreigners": Competing categorizations within gold mining spaces in Guinea. *Resources Policy* 40: 117–127.

Botchway F.N. (1995) Pre-colonial methods of gold mining and environmental protection in Ghana. *Journal of Energy and Natural Resources Law* 13 (4): 299–311.

Brightman M. and Lewis J. (2017) *The anthropology of sustainability. Beyond development and progress*, London: Palgrave Macmillan.

Childs J. (2008) Reforming small-scale mining in sub-Saharan Africa: Political and ideological challenges to a fair trade gold initiative. *Resources Policy* 33: 203–209.

Childs J. (2014) A new means of governing artisanal and small-scale mining? Fair trade gold and development in Tanzania. *Resources Policy* 40: 128–136.

Chowdhury N.S. (2016) Mines and signs: Resource and political futures in Bangladesh. *Journal of the Royal Anthropological Institute* 22 (S1): 87–107.

Cooke B. and Kothari U. (2001) *Participation: The new tyranny?*, London: Zed Books.

Corrigan C. (2014) Breaking the resource curse: Transparency in the natural resource sector and the extractive industries transparency initiative. *Resources Policy* 40: 17–30.

Coumans C. (2011) Occupying spaces created by conflict: Anthropologists, development NGOs, responsible investment, and mining. *Current Anthropology* 52 (S3): S29–43.

Cross J. (2011) Detachment as a corporate ethic: Materializing CSR in the diamond supply chain. *Focaal* 60: 34–46.

Cuvelier J. (2017) *Leaving the beaten track? The EU regulation on conflict minerals*, Africa Policy Brief n. 20, EGMONT Royal Institute for International Relations.

Cuvelier J., Van Bockstael S., Vlassenroot K. and Iguma C. (2014) *Analyzing the impact of the Dodd-Frank Act on Congolese livelihoods*, SSRC Conflict Prevention and Peace Forum.

D'Angelo L. (2019) Diamonds and plural temporalities. Articulating encounters in the mines of Sierra Leone. In: Pijpers R.J. and Eriksen T.H. (eds) *Mining encounters. Extractive industries in an overheated world*. London: Pluto Press, 138–155.

D'Angelo L., and Pijpers R.J. (2018) Mining temporalities: An overview. *The Extractive Industries and Society* 5: 215–222.

Dolan C. and Rajak D. (2011) Introduction: Ethnographies of corporate ethicizing. *Focaal* 60: 3–8.

Dolan C. and Rajak D. (2016) *The anthropology of corporate social responsibility*, New York and Oxford: Berghahn.

Dougherty M.L. (2019) Boom times for technocrats? How environmental consulting companies shape mining governance. *The Extractive Industries and Society* 6: 443–453.

DuPisani J. (2006) Sustainable development: Historical roots of the concept. *Environmental Sciences* 3 (2): 83–96.

Ferguson J. (1990) *The anti-politics machine. "Development", depoliticization, and bureaucratic power in Lesotho*, Cambridge: Cambridge University Press.

Ferguson J. (1999) *Expectations of modernity. Myths and meanings of urban life on the Zambian Copperbelt*, Berkeley: University of California Press.

Ferry E.E. and Limbert M.E. (2008) Introduction. In: Ferry E.E. and Limbert M.E. (eds) *Timely assets: The politics of resources and their temporalities*. Santa Fe: School for Advanced Research Press, 3–24.

Filer C. and Le Meur P.Y. (2017) *Large-scale mines and local-level politics. Between New Caledonia and Papua New Guinea*, Canberra: ANU Press.

Fisher E. (2018a) Fair trade movement. In: Callan H. (ed.) *The international encyclopaedia of anthropology*. Wiley online.

Fisher E. (2018b) Solidarities at a distance: Extending Fairtrade gold to east Africa. *The Extractive Industries and Society* 5: 81–90.

Fisher E. and Childs J. (2014) An ethical turn in African mining: Voluntary regulation through fair trade. In: Bryceson D.F., Fisher E., Jønsson J.B. and Mwaipopo R. (eds) *Mining and social transformation in Africa*. London and New York: Routledge, 130–147.

Fisher E., Luning S., de Theije M., D'Angelo L., Pijpers R., Massaro L. et al. (2021) Transforming matters: Sustaining gold lifeways in small-scale mining. *Current Opinion in Environmental Sustainability* 49: 190–200.

Fraser J. (2019) Creating shared value as a business strategy for mining to advance the United Nations Sustainable Development Goals. *The Extractive Industries and Society* 6 (3): 788–791.

Fridell G. (2006) Fair trade and neoliberalism. Assessing emerging perspectives. *Latin American Perspectives* 33 (6): 8–28.

Gardner K., Ahmed Z., Bashir F. and Rana M. (2012) Elusive partnerships: Gas extraction and CSR in Bangladesh. *Resources Policy* 37: 168–174.

Gilberthorpe E. (2007) Fasu solidarity: A case study of kin networks, land tenure, and oil extraction in Kutubu, Papua New Guinea. *American Anthropologist* 109 (1): 101–112.

Goodman M.K. (2004) Reading fair trade: Political ecological imaginary and the moral economy of fair trade foods. *Political Geography* 23: 891–915.

Günel G. (2019) *Spaceship in the desert. Energy, climate change, and urban design in Abu Dhabi,* Durham and London: Duke University Press.

Halvaksz J.A. (2008) Whose closure? Appearances, temporality, and mineral extraction in Papua New Guinea. *Journal of the Royal Anthropological Institute* 14: 21–37.

Han Onn A. and Woodley A. (2014) A discourse analysis on how the sustainability agenda is defined within the mining industry. *Journal of Cleaner Production* 84: 116–127.

Hilson G. (2008) "Fair trade gold": Antecedents, prospects and challenges. *Geoforum* 39: 386–400.

Hilson G., Hilson A. and McQuilken J. (2016) Ethical minerals: Fairer trade for whom? *Resources Policy* 49: 232–247.

Hopwood B., Mellor M. and O'Brien G. (2005) Sustainable development: Mapping different approaches. *Sustainable Development* 13 (1): 38–52.

Jenkins H. and Yakovleva N. (2006) Corporate social responsibility in the mining industry: Exploring trends in social and environmental disclosure. *Journal of Cleaner Production* 14: 271–284.

Joyce S.A. and MacFarlane M. (2001) *Social impact assessment in the mining industry: Current situation and future directions,* MMSD Report n. 46, London.

Kapelus P. (2002) Mining, corporate social responsibility and the "community": The case of Rio Tinto, Richards Bay Minerals and the Mbonambi. *Journal of Business Ethics* 39: 275–296.

Kemp D., Owen J.R. and van de Graaff S. (2012) Corporate social responsibility, mining and "audit culture". *Journal of Cleaner Production* 24: 1–10.

Kirsch S. (2010) Sustainable mining. *Dialectical Anthropology* 34: 87–93.

Kirsch S. (2014) *Mining capitalism: The relationship between corporations and their critics,* Oakland: University of California Press.

Knierzinger J. and Sopelle I.T. (2019) Mine closure from below: Transformative movements in two shrinking West African mining towns. *The Extractive Industries and Society* 6: 145–153.

Lanzano C. (2018) Gold digging and the politics of time: Changing timescapes of artisanal mining in West Africa. *The Extractive Industries and Society* 5: 253–259.

Lanzano C. and Arnaldi di Balme L. (2017) Des "puits burkinabè" en Haute Guinée: Processus et enjeux de la circulation de savoirs techniques dans le secteur minier artisanal. *Autrepart* 82: 87–108.

Lawrence R. and Kløcker Larsen R. (2017) The politics of planning: Assessing the impacts of mining on Sami lands. *Third World Quarterly* 38 (5): 1164–1180.

Lélé S. (1991) Sustainable development: A critical review. *World Development* 19 (6): 607–621.

Le Meur P.Y. (2015) Anthropology and the mining arena in New Caledonia: Issues and positionalities. *Anthropological Forum* 25 (4): 405–427.

Li F. (2009) Documenting accountability: Environmental impact assessment in a Peruvian mining project. *PoLAR* 32 (2): 218–236.

Li F. (2011) Engineering responsibility: Environmental mitigation and the limits of commensuration in a Chilean mining project. *Focaal* 60: 61–73.

Li F. (2015) *Unearthing conflict. Corporate mining, activism, and expertise in Peru*, Durham and London: Duke University Press.

Limbert M.E. (2010) *In the time of oil. Piety, memory, and social life in an Omani town*, Stanford: Stanford University Press.

Lodhia S. and Hess N. (2014) Sustainability accounting and reporting in the mining industry: Current literature and directions for future research. *Journal of Cleaner Production* 84: 43–50.

Loher D. (2020) Everyday suffering and the abstract time-reckoning of law. Reflections on the allocation of responsibility for an asbestos disaster in Italy. *Journal of Legal Anthropology* 4 (2): 17–38.

Lozano R. (2008) Envisioning sustainability three-dimensionally. *Journal of Cleaner Production* 16: 1838–1846.

Luning S. (2010) Beyond the pale of property: Gold miners meddling with mountains. In: Panella C. (ed.) *Worlds of debts. Interdisciplinary perspectives on gold mining in West Africa*. Amsterdam: Rozemberg, 25–48.

Luning S. (2013) Gold from "rock to ring". *Etnofoor* 25 (2): 159–174.

Luning S. (2018) Mining temporalities: Future perspectives. *The Extractive Industries and Society* 5: 281–286.

Luning S. and de Theije M. (2014). Global gold connections. Ethical consumption and the beauty of bonding artisans. In: Barendrecht B. and Jaffe R. (eds) *Green consumption. The global rise of eco-chic*. London, New Delhi, New York and Sydney: Bloomsbury, 56–70.

Marston A. (2013) Justice for all? Material and semiotic impacts of Fair Trade craft certification. *Geoforum* 44: 162–169.

Mazzeo A. (2021) *Dust inside. Fighting and living with asbestos-related disasters in Brazil*, New York: Berghahn.

Mebratu D. (1998) Sustainability and sustainable development: Historical and conceptual review. *Environmental Impact Assessment Review* 18 (6): 493–520.

MMSD (2002) *Breaking new ground. Mining, minerals and sustainable development*, London: Earthscan.

Moore H. (2017) What can sustainability do for anthropology? In: Brightman M. and Lewis J. (eds) *The anthropology of sustainability. Beyond development and progress*. London: Palgrave Macmillan, 67–80.

Morrison-Saunders A. and Arts J. (2004) *Assessing impact. Handbook of EIA and SEA follow-up*, London: Earthscan.

Mususa P. (2010) Contesting illegality: Women in the informal copper business. In: Fraser A. and Larmer M. (eds) *Zambia, mining, and neoliberalism. Boom and bust on the globalized Copperbelt*. New York: Palgrave Macmillan, 185–208.

Nielsen K.B. (2017) Unclean slates: Greenfield development, land dispossession and "EIA struggles" in Goa. *South Asia: Journal of South Asian Studies* 40 (4): 844–861.

Olivier de Sardan J.P. (1993) Le développement comme champ politique local. *Bulletin de l'APAD* 6. DOI: doi:10.4000/apad.2473.

Olofsson T. (2020) Imagined futures in mineral exploration. *Journal of Cultural Economy* 13 (3): 265–277.

Owen J.R. and Kemp D. (2013) Social licence and mining: A critical perspective. *Resources Policy* 38: 29–35.

Pels P. (2015) Modern times. Seven steps toward an anthropology of the future. *Current Anthropology* 56 (6): 779–796.

Perrault T. (2015) Performing participation: Mining, power, and the limits of public consultation in Bolivia. *The Journal of Latin American and Caribbean Anthropology* 20 (3): 433–451.

Persoon G.A. and van Est D.M.E. (2000) The study of the future in anthropology in relation to the sustainability debate. *Focaal* 35: 7–28.

Phadke R. (2018) Green energy futures: Responsible mining on Minnesota's Iron Range. *Energy Research & Social Science* 35: 163–173.

Pijpers R.J. (2016) Mining, expectations and turbulent times: Locating accelerated change in rural Sierra Leone. *History and Anthropology* 27 (5): 504–520.

Pijpers R.J. (2017) I want to be a millionaire: Survival, trust and deception in Sierra Leone's economy of dreams. In: de Rooij V. (ed.) *Anthropological knowledge on the move*. Amsterdam: Amsterdam University Press, 135–153.

Power T. (2002) *Digging to development? A historical look at mining and economic development*, Oxford: Oxfam.

Purvis B., Mao Y. and Robinson D. (2019) Three pillars of sustainability: In search of conceptual origins. *Sustainability Science* 14: 681–695.

Pusceddu A.M. (2020) *Energopolitics of transition: Resource making and contestation in the Portuguese lithium rush*. Paper presented at the 16th EASA Biennial Conference, Lisbon (unpublished).

Rahnema M. (1992) Participation. In: Sachs W. (ed.) *The development dictionary*. London: Zed Books, 116–131.

Rajak D. (2011) *In good company. An anatomy of corporate social responsibility*, Stanford: Stanford University Press.

Robinson J. (2004) Squaring the circle? Some thoughts on the idea of sustainable development. *Ecological Economics* 8 (4): 369–384.

Rogers D. (2012) The materiality of the corporation: Oil, gas, and corporate social technologies in the remaking of a Russian region. *American Ethnologist* 39 (2): 284–296.

Sandlos J. and Keeling A. (2016) Aboriginal communities, traditional knowledge, and the environmental legacies of extractive development in Canada. *The Extractive Industries and Society* 3: 278–287.

Sarmiento M., Ayala H., Urán A., Giraldo B., Perea J. and Mosquera A. (2013) Legitimidad e innovación en la minería: el caso del Programa Oro Verde. *Letras Verdes. Revista Latinoamericana de Estudios Socioambientales* 14: 284–303.

Sharp J. (2006) Corporate social responsibility and development: An anthropological perspective. *Development Southern Africa* 23 (2): 213–222.

Smith J.H. (2011) Tantalus in the digital age: Coltan ore, temporal dispossession, and "movement" in the Eastern Democratic Republic of the Congo. *American Ethnologist* 38 (1): 17–35.

Smith S.M., Shepherd D.D. and Dorward P.T. (2012) Perspectives on community representation within the Extractive Industries Transparency Initiative: Experiences from south-east Madagascar. *Resources Policy* 37: 241–250.

Spiegel S. (2017) EIAs, power and political ecology: Situating resource struggles and the techno-politics of small-scale mining. *Geoforum* 87: 95–107.

Sydow J. (2016) Global concepts in local contexts: CSR as "anti-politics machine" in the extractive sector in Ghana and Peru. In: Dolan C. and Rajak D. (eds) *The anthropology of corporate social responsibility*. New York and Oxford: Berghahn, 217–242.

Szolucha A. (2018) Anticipating fracking: Shale gas developments and the politics of time in Lancashire, UK. *The Extractive Industries and Society* 5: 348–355.

Tarus L., Hufford M. and Taylor B. (2017) A green new deal for Appalachia: Economic transition, coal reclamation costs, bottom-up policymaking. *Journal of Appalachian Studies* 23 (2): 151–169.

Teixeira R.O.S., Zhouri A. and Dias Motta L. (2021) Os estudos de impacto ambiental e a economia de visibilidades do desenvolvimento. *Revista Brasileira de Ciências Sociais* 36 (105).

UN (2002) *Report of the World Summit on Sustainable Development, Johannesburg, 26 August–4 September 2002*, New York: United Nations.

Van Bockstael S. (2018) The emergence of conflict-free, ethical, and Fair Trade mineral supply chain certification systems: A brief introduction. *The Extractive Industries and Society* 5: 52–55.

Vogel C. and Raeymaekers T. (2016) Terr(it)or(ies) of peace? The Congolese mining frontier and the fight against "conflict minerals". *Antipode* 48 (4) 1102–1121.

Walsh A. (2012) After the rush: Living with uncertainty in a Malagasy mining town. *Africa* 82 (2): 235–251.

Warnaars X. (2012) Why be poor when we can be rich? Constructing responsible mining in El Pangui, Ecuador. *Resources Policy* 37: 223–232.

Watts M. (2005) Righteous oil? Human rights, the oil complex, and corporate social responsibility. *Annual Review of Environment and Resources* 30: 373–407.

WB (2004) *Striking a better balance: The World Bank Group and extractive industries – the final report of the extractive industries review*, Washington D.C.: World Bank.

WCED (1987) *Our common future*, Oxford: Oxford University Press.

Welker M. (2014) *Enacting the corporation. An American mining firm in post-authoritarian Indonesia*, Berkeley: University of California Press.

Westman C.N. (2013) Social impact assessment and the anthropology of the future in Canada's tar sands. *Human Organization* 72 (2): 111–120.

Weszkalnys G. (2015) Geology, potentiality, speculation: On the indeterminacy of first oil. *Cultural Anthropology* 30 (4): 611–639.

Whitmore A. (2006) The emperor's new clothes: Sustainable mining? *Journal of Cleaner Production* 14: 309–314.

Wiegink N. (2018) Imagining booms and busts: Conflicting temporalities and the extraction–"development" nexus in Mozambique. *The Extractive Industries and Society* 5: 245–252.

10

TECHNOLOGY

Lorenzo D'Angelo

1 Introduction

In the mid-1980s, anthropologist Ricardo Godoy complained about the lack of interest exhibited by his colleagues in both the economy and the production processes of mining (Godoy 1985a; cf. Rubbers this volume). This lack of interest seems to have been accompanied by limited attention paid, until recently, to anything that can be defined as technology (Ingold 2000; Pfaffenberger 1988, 1992). Often, technology is regarded as a simple "making and using" that does not deserve particular attention (Winner 1986). At the most, it is a question of acknowledging that "tools can be 'used well or poorly' and for 'good or bad purposes'" (Winner 1986: 6). Thus, Brian Pfaffenberger (1988) is not surprised that in the studies on mining too, "technology is only rarely seen [...] as a subject that is itself intrinsically of interest" (ibid.: 236). For this reason, he agrees with Godoy (1985a) when the latter complains that in monographs the analysis of the "productive" aspects is usually squeezed into one or two pages. And, as far as mining technology is concerned, the analysis is often reduced to a mere list of objects, with details flattened to the descriptive level of the material culture (Godoy 1985a: 211; Pfaffenberger 1988: 236).

Since Godoy and Pfaffenberger published their reviews on the anthropology of mining and technology, respectively, the anthropological literature has branched out and been enriched by new perspectives thanks to the interdisciplinary contributions offered by Science and Technology Studies (STS) (e.g. Bijker et al. 2012), Feminist Technology Studies (e.g. Bray 2007), the renewed interest in the Francophone Tradition of Anthropology of Techniques (Naji and Douny 2009), and the individual works of anthropologists such as Bruno Latour and his Actor Network Theory (e.g. Latour 2005); Tim Ingold and his "Spider's perspective" (e.g. Ingold 2011), Alf Hornborg's reflections on "machine fetishism" (e.g. 2013), as well as the contributions of Pfaffenberger himself (e.g. 1988, 1992, 2002).

DOI: 10.4324/9781003018018-10

Despite this theoretical richness, technological issues do not seem to have made inroads among anthropologists interested in mining or natural resource extraction (see also Massaro and de Theije 2018). Nevertheless, there are signs of a recent shift which this chapter intends to highlight by intertwining studies of various kinds, largely but not only by anthropologists. Given the complexity of the issues and the dynamism of the literature on technology, interdisciplinary exchanges and loans appear inevitable as well as advantageous, especially from a synchronic and diachronic comparative perspective. For its part, anthropological analysis offers a specific contribution by socially and culturally contextualizing meanings and practices with a holistic view capable of grasping the perspective of social actors and the ways in which they interact or incorporate mining technologies and their continuous changes (cf. Pfaffenberger 1988: 245).

With the notion of a "sociotechnical system", Pfaffenberger (1988) offers a good starting point to think about mining technology. With this notion Pfaffenberger intends to emphasize, from a symbolic anthropological perspective, the essentially systemic and social (non-technical) nature of technology (ibid.: 241). In other words, for Pfaffenberger technology is not simply a set of objects that belong to the material culture of a given society. It is also a form of "humanized nature", or better still, to quote Marcel Mauss, a "total social fact", that is, a "fact" which has material, social, political, and symbolic dimensions (ibid.: 244, 249).

The Maussian notion of social fact is not far from Bruno Latour's notion of assemblage (Latour 2005; see Kasuga 2010). Indeed, it could be argued that mining technologies are, in many respects, assemblages of heterogeneous elements – tools, knowledge, body skills, and extractive methods – that articulate and produce sociomaterial syntheses adapted and influenced by specific ecological, material, historical, political, and cultural circumstances.[1] Taking a further step in following Marcel Mauss's synthetic view, it is also possible to extend the definition of technology as an effective action on matter (Lemonnier 1996) to also include its ability to act upon and shape human subjectivities (Warnier 2009).

These three definitions – technology as sociotechnical system, as assemblage, and as effective action on matter and subjects – are useful to circumscribe the scope of this chapter. The next section highlights how mining technology is the result of creative assemblages of heterogeneous elements, technical and non-technical, which move and adapt to different places, times, and contingent circumstances. It also demonstrates that technological changes can have social effects on local social structures, gender, and generational dynamics. Against the background of a definition of technology as effective action on matter and subjects, the third section examines a central issue for miners, namely, the efficiency of the techniques and tools used to extract natural resources. This section also highlights some paradoxical aspects related to this efficiency, aspects that warn us to avoid linear and simplistic views on so-called "technological progress". It also stresses the importance of producing a detailed analysis of each context under consideration. The fourth section focuses on miners' knowledge and technical skills, on gender and on the bodily aspects of the relationship with mining technologies. The chapter concludes with some consideration of the most recent technological developments and suggests potential future directions for anthropological research on mining technology.

2 Mobility, mining frontiers, and social change

Mining technology is often examined in terms of inventions, which are in turn associated with progress. However, what appears to be "new" may simply be a rediscovery – a forgotten and then resumed practice – or the readjustment of tools or techniques that already existed in the past or elsewhere. Indeed, exchanges and loans fuel innovations which are often due not only to expert engineers but also to unknown miners. Often, these innovations are linked to the presence of migrants – free and unfree labourers – who, while travelling, bring with them knowledge, tools, and experiences. In this context, a dialectical exchange between archaeology, history, and anthropology seems to be a necessary step to avoid the temptation of considering technology only in terms of mere inventions. On the contrary, it is a matter of showing how mining technology is often the result of continuous processes of diffusion, adaptation, and re-invention.

Russell-Wood (1977) notes that the European miners who, in the seventeenth century, managed the gold mines in Minas Gerais in Brazil, had no specific experience in this sector. Consequently, they depended almost entirely on the knowledge and experience possessed by the slaves who worked for them. Many of these slaves came from West African regions with a long history of both metallurgy and gold mining, such as present-day Ghana, formerly known as the Gold Coast for its gold mines. In this African context, gold mining began well before the arrival of European mining companies (e.g. Pijpers 2020). Already in pre-colonial times, the Gold Coast's Obuasi mines attracted migrant workers and required the use of slaves imported from other territories to the north. These free and unfree workers brought with them methods, technologies, and experiences accumulated elsewhere – experiences that were transferred, at least in part, to Brazil as well (Ofosu-Mensah 2017).

The role of migrants in the innovation and dissemination of mining techniques and technologies is also addressed by Raymond Dumett in a book that is already a classic, *El Dorado in West Africa* (1998). The notion of "mining frontier" plays an important role in this book. Dumett's effort is to elaborate this notion in non-Eurocentric terms. In this light, one of its main goals is to show how the mining technology adopted by African peasant-miners also contributed to the expansion of the colonial frontier of gold exploitation in the Gold Coast both before and during the development of mining capitalism in the late nineteenth century. These miners, in fact, adopted continuous innovations, such as "fire setting", a technique for breaking rocks adopted by the Nzema people who worked seasonally in the deep-reef gold mines of the Akan gold mining states.

Dumett's reflection on the mining frontier has inspired other scholars, in particular, those with an interest in the relationship between mobility and mining technology (e.g. Werthmann and Grätz 2012). Ross (2014) uses this notion to examine some important innovations in the tin mining industry in Malaysia, which has benefited from technologies employed in completely different contexts. For example, the introduction of *chin-chia* – a wooden bucket-chain mechanism used by the Chinese community in the Malay peninsula in the mid-nineteenth century to collect water for domestic use – proved particularly useful in the open pits. The use of this mechanism, in fact, made possible the removal of water from the pits. In turn this allowed miners to dig deeper, thus recovering greater quantities of tin.

Ross (2014) also highlights how the continued growth in international demand for tin and the need to exploit deposits with a low metal content stimulated the introduction of other technologies. These innovations did not entirely replace the previous ones. Between the end of the nineteenth century and the beginning of the twentieth, Chinese extractive techniques coexisted with the hydraulic technology imported from and used in California and Australia between 1850 and 1860 to extract gold. This technology, in turn, was nothing more than the adaptation of hydraulic techniques already known to the Romans when they occupied the Iberian Peninsula (Isenberg 2005). As in the case of gold mining in North America, in Malaysia too the water of the rivers was pressurized to wash entire hills from top to bottom. In this way, European tin mining companies were able to exploit deposits that were less accessible to and productive with panning technologies. However, they left behind barren and desolate landscapes (Ross 2014).

The notion of mining frontier is also considered by Lanzano (2020) to examine the technological changes in two different gold mining areas of Guinea Conakry and in neighbouring Burkina Faso. At the centre of this comparison are the Burkina Faso miners: on the one hand, those who migrate to Guinea Conakry to extract gold from primary hard-rock deposits by digging the so-called "Burkinabe pits" and, on the other hand, those who remain in Burkina Faso and extract gold from by-products using cyanide-based processing techniques. In the wake of Sabine Luning's critical reflection on the notion of frontier – which considers the risk of thinking of the areas on the margins of mining capitalism expansion in terms of *terra nullius* (Luning 2018; see also Golub 2019) – Lanzano underlines the "creative process of (often) selective appropriation" of technological innovations in local contexts (2020: 258). Following Pfaffenberger (1992), he further emphasizes the embodied and not purely technical nature of the technological: technological transformations can produce or influence social changes and vice versa.

From different theoretical perspectives, the works of Verbrugge and Van Wolputte (2015), Dessertine (2016), and Calkins (2016) are further examples of how the introduction of a specific technology, in their case the metal detector, can bring about social changes that affect the relationships between genders and generations. Verbrugge and Van Wolputte's (2015) study focuses on women seeking artisanal gold in Makongolosi in south-west Tanzania. In this highly competitive and male-dominated context, women try to carve out their own room for action. The difficulties, however, are many. Men are allowed to dig deep; they can use expensive metal detectors through which they can find gold in the form of nuggets that can be sold at a good price, allowing high earnings. Women, on the other hand, usually work alone or in small groups in abandoned mines far from their villages. They collect the gold that is found in small quantities after the gold rocks have been pulverized with ball mills. Like the metal detectors, ball mills have been introduced recently, and they are owned by male miners. When female miners pay male miners to operate these machines, they feel exploited and duped. In fact, the owners of the ball mills not only keep prices high but resell the remains of the tailings to mining companies capable of extracting the tiny quantities of gold left in them. In so doing, they gain extra income that is not shared with the women who brought the stones. Similarly, women know they can be easily fooled by male operators who rent their metal detectors. The latter use headphones to listen to the signals emitted by this machinery and "may or may not lie about [their] findings by keeping the information for [them]self, and come back later to dig the soil" (ibid.: 182). Considering the circle of dependency triggered or reinforced by these technologies, Verbrugge and Van Wolputte (2015) conclude that they are effectively "agents of poverty" (Hilson and Pardie 2006) that reinforce or reproduce social and economic inequalities between genders.

Dessertine's work (2016) also examines a situation of marginalization of women miners associated with the introduction of the metal detector. Her analysis is based on the comparison between two different periods of field research carried out in north-eastern Guinea Conakry. She observes that during the dry season of 2011, the gold mines were populated with artisanal miners – men and women in equal measure – from different villages. These people lived in camps that apparently looked like miniature villages but which, in reality, had specific and unique dynamics that made them a sort of heteropic space (see Werthmann 2010). In these spaces, women were allowed, for example, to go to bars or have "short-term marriages". Furthermore, the prevailing form of income distribution was based on egalitarian principles. With the introduction of the metal detector in 2012, Dessertine notes that the mobility of miners increased significantly. Following their metal detectors, in fact, the miners had no reasons to spend long periods of time in the same place. As a result, camps were not formed; miners returned to their villages with greater regularity and frequency, flaunting and squandering, more than they did before, the small or large amounts of wealth accumulated by selling gold. As Dessertine notes, in the mining region she examined "a female presence is characterised by a prolonged stay in a single place, with only brief, local absences" (2016: 439). Thus, the spread of the metal detector helped to trigger a social dynamic dominated once again by men in which women, as in the case of southwest Tanzania, find themselves at the margins.

The metal detector is also at the centre of ethnographic research conducted by Calkins (2016) in north-eastern Sudan. As in previous cases, the metal detector was a recent innovation when Calkins carried out her fieldwork between 2009 and 2010. Here, the basic issue is a little bit different: understanding how mining technology contributes to resource-making (see also Akiwumi and D'Angelo 2018; de Rijke 2018; Kama 2020). More precisely, Calkins examines how the metal detector used by Sudanese miners contributes to the emergence of a specific local mineralogical category, the so-called "clean gold", which is a type of gold recoverable in the most superficial layers of soil and which has a high degree of purity. Calkins is careful to avoid making linear connections between technological and social changes. In this perspective, she considers the metal detector as part of a wider techno-economic infrastructure characterized by fluidity and openness rather than the stability and rigidity. Moreover, Calkins's attention focuses on the relationship between generations rather than between genders and on how the infrastructure of which the metal detector is a part brings out a moral order based on luck and individual profit – an order that questions the moral and economic authority of the elderly, which is based on hierarchical relationships of dependence.

Beyond the specific contexts and theoretical approaches of each of these studies, this section has highlighted the particular interest shown by anthropologists and historians in the question of the relationship between technology and social change. What unites these different approaches is the search for specific modalities of local interaction with technologies spread across different times and places. The next section focuses on technological changes intended as the search for efficiency (and sustainability). As we shall see, these studies must deal with controversial issues that can have paradoxical implications including the question of whether large-scale technologies are more efficient than artisanal ones.

3 Efficiency and sustainability of artisanal and large-scale mining

The efficiency of extraction methods and technologies is a central issue for miners, especially when the deposits are poor at the start or become impoverished after long periods of extraction. In this regard, the technologies of large-scale mining companies are usually appreciated by experts when compared to the artisanal sector, as they are organized according to standardized industrial work models and are therefore seemingly more efficient, profitable, and respectful of the environment. In contrast, technologies used at artisanal or small-scale levels – especially when they are carried out illegally – are portrayed as rudimentary, inefficient, and ecologically harmful (see, for example, Hilson and Van der Vorst 2002; Seccatore and de Theije 2017; Sinding 2005), as well as dangerous to human health (e.g. Maia et al. 2019; Maponga and Ruzive 2002; Veiga et al. 2014).

In the wake of the debate on the efficiency and sustainability of artisanal mining (on the issue of sustainability, see Lanzano this volume), Massaro and de Theije (2018) examine the efforts of a mining cooperative in the state of Mato Grosso in Brazil that uses techniques and technologies considered ecologically more sustainable than those used by the so-called *garimpos*, that is, the illegal miners. Their study argues that the introduction of relatively new mining technologies (i.e., the exploration drill) and extractive techniques (i.e., the cyanidation process) – encouraged above all by international agencies, NGOs, and the government – can only work if the miners are involved in the same process of change. To regard these miners as inexperienced workers who are unable to plan their actions for the future is to ignore their ingenious solutions, their capacity to improve these same technologies. In other words, to underestimate miners' agency means exposing these projects for mining sustainability to the risk of failure.

Some anthropological, historical, and economic studies, however, have radically questioned the idea that artisanal mining is less efficient and less sustainable than large-scale mining (e.g. Godoy 1985b). There are two main arguments here. Firstly, since the colonial era the narrative of the primitive, inefficient, and environmentally damaging (usually "illegal") artisanal miner has favoured, on the one hand, the granting of mining licenses to mining companies and, on the other, the marginalization and, in some cases, the criminalization of the same artisanal miners (e.g. Thabane 2003). Secondly, several scholars point out that, under certain conditions, the artisanal or small-scale extraction of certain types of resources (e.g. alluvial gold or diamonds) can be as efficient, or even more efficient, than large-scale mining (e.g. Curtin 1973; Dumett 1998). The efficiency of small-scale extraction clearly emerges if one considers the amount of mineral recovered in relation to the costs and efforts put to get it, or the number of high-quality gems recovered in relation to the amount lost during the treatment operations – a fact that some colonial geologists and mining experts reluctantly admitted also in their reports (e.g. d'Avignon 2018; Greenhalgh 1985) and some mining engineers have recently recognized with simple testing methods (Teschner et al. 2017). Not only that, but the alleged superiority of some mining companies is often based on the exploitation of, so to speak, "traditional" extractive and organizational techniques. Where mining companies have relied exclusively on highly capitalized and mechanized systems, they have encountered enormous difficulties, so much so that many of them have failed (e.g. Ross 2014; Silver 1981). Some large-scale companies have cynically treasured these failed experiences. For this reason, more and more, they let artisanal or small-scale miners literally do the dirty work, delegating to them the risks of explorative operations and the economic and social problems that derive from extractive activities (Luning 2014; Verbrugge 2020). Of course, this situation does not exclude the possibility that subcontracted small-scale miners can benefit from arrangements with mining companies, especially when the latter are not permitted to mine themselves (Fisher 2008).

Among the failed experiences, it is worth lingering on the case of diamond mining in Sierra Leone. In the early 1960s, the Diamond Exploration Company decided to buy a mining license from the government of Sierra Leone to explore a stretch of the Sewa River that had proved extremely productive for local artisanal miners. Confident of being able to make far greater profits than local miners thanks to the financial capital and technology at its disposal, the company imported a powerful – and very expensive – dredge which was capable of extracting and sifting the gravel from the bottom of the river, day and night, in any weather condition. Despite these premises, the project was abandoned after a few years as the total value of the stones recovered from the dredge was never able to cover the costs of the personnel and maintenance of the dredge (D'Angelo 2018).

This example points toward an apparently obvious conclusion. As Thoburn (1977) notes, there are no universally effective mining technologies. Each technological assemblage must take into account the specific materiality of the extracted resources (see Ferry this volume) and the contexts in which they are introduced and adapted. In certain circumstances, it may happen that a tool or an instrument deemed to be inefficient is, in reality, the most suitable for the type of resource extracted or the needs of the miners. These apparently paradoxical circumstances require careful analysis of each context. To understand, for example, how much the legal context can affect technological choices, it is worth returning to the Sierra Leonean case by comparing the diamond mines of Lesotho in the 1960s and 1970s (Thabane 2003) and those of Sierra Leone in the 1950s and 1960s (D'Angelo 2019). In both cases, some of the illegal artisanal miners used metal containers similar to basins or perforated pots as sieves. From the point of view of mining engineers, these sieves were a clear demonstration of the inefficiency of artisanal technologies. This negative judgement, later partially retracted, did not take into account the conditions of illegal work and the practical objectives of the miners. From the point of view of an illegal diamond seeker who worked covertly – often, as in the case of Sierra Leone, in the darkest hours of the night – the priority was to retrieve the largest gems in the shortest possible time. Searching for the smaller stones, of little commercial value in the illegal market, meant wasting time and increasing the risk of being caught red-handed by the police or the security forces that regularly patrolled the mining areas. Moreover, compared to an industrially made sieve, a perforated pot is easier to disguise and transport without attracting the attention of the authorities. By interviewing the Lesotho miners, and comparing some subsequent engineering reports, Thabane (2003) also discovered that the holes created by these workers with variable-sized awls were not random. The smaller holes were placed in the centre of the bowl floors, with the larger ones on the sides. This arrangement follows the gravitational logic that pushes stones with the highest specific gravity such as diamonds to the bottom and larger, useless stones to the sides. This simple and ingenious solution is anything but coarse if one considers the aims and the means available to these miners.

As this example shows, the question of efficiency is not a purely technical one. In the context of diamond mining in Lesotho and Sierra Leone, accusing artisanal miners of using inefficient technologies was a way of justifying their marginalization in favour of large-scale companies. Thus, the question of efficiency and technological development must also be thought in terms of political objectives. Historically, the development of mechanization and the search for efficient technical solutions has often had the primary purpose not only of maximizing the ratio between costs and earnings, but also of reducing the dependence of capital on labour.[2] On equal terms, reducing the need for labour means, in fact, increasing unemployment and reducing wages, as well as weakening the ability of workers to organize themselves, resist, and perhaps strike (e.g. Thoburn 1977).

The search for technological solutions, which can be considered efficient from an engineering point of view – what is often considered as a sign of technological "progress" or "development" – finally, and paradoxically, raises questions of environmental sustainability. Large-scale technologies, precisely because of their extreme "efficiency" in terms of high productive capacity and absolute volumes

processed, can have disastrous long-term environmental effects that are no less significant – in many cases, certainly, much more relevant – than those usually
attributed to artisanal miners (e.g. Isenberg 2005; see also Manuel 2017).

In short, all these examples show that technological efficiency cannot be calculated with a simple formula that abstracts from the context in which this same
technology is used. Thinking of mining technologies purely in terms of engineering efficiency is a mistake as it is simplistic to think that bigger is (necessarily)
better. At the same time, it would be misleading to think of artisanal and large-
scale mining as opposing and incompatible extractive levels and modalities of
extracting resources even if the reality on the ground often reveals the tension
between the two on access to land and deposits. The history of mining technology
is often made up of additions rather than replacements or linear transitions. These
additions have, in turn, meant finding or creating the conditions for the coexistence (Luning and Pijpers 2017), or the articulation (D'Angelo 2018) of different
modes of extraction.

4 Bodies: skills, gender, race, and power

Keeping in mind the definition of technology as effective action on objects and
subjects (Warnier 2009), it becomes apparent that technological changes affect not
only what miners do, and how they do it, but also what they think of their profession and themselves, of their class, race, and gender identity. In her study of the
marble quarrymen of Carrara in Italy, Leitch (1996) argues that working in the
mine can be considered as an incorporated activity that the quarrymen themselves
compare to a form of art rather than a simple job. This art requires skills, both
individual and collective, transmitted from generation to generation, or rather,
from father to son. The Italian marble quarries are in fact the domain of masculinity. The confidence these workers have in their knowledge and skills, as well as in
their machinery and tools, translates into a self-representation that sees them as men
in a battle with the mountain, "agents in the transformation of the natural world"
(ibid.: 245).

As Rebecca Scott shows in *Removing Mountains*, mining technology is also
strongly associated with masculinity in West Virginia (Scott 2010). Here, the feminine nature of the Appalachian mountains is tamed with dynamite and mammoth
draglines that flatten their profiles day after day. Despite the concerns of environmentalists, mountaintop removal is presented by major mining companies as a
way of improving the landscape, a necessary premise for the progress and economic
development of a poor region. Scott points out that, in the culturally hyper-mas-
culinized context of the American coal industry, miners tend to identify with their
machinery, to the point that, underground, "both are called 'miners'" (ibid.: 81).
However, the technology of mountaintop removal operates above all on the surface and its high degree of automation makes skilled labour less and less necessary.
But instead of undermining the symbolic assumptions on which the masculine
narrative of the coal miners is based, paradoxically, this technology helps to

intensify the identification of male workers with their equipment (see also Saugeres 2002). The identification is so powerful that women "are constructed as lacking in technical expertise", and consequently their work is diminished (Scott 2010: 81).

Anthropologist Jessica Smith Rolston, who worked as a driver and plant technician in the Wyoming coal mines, the same mines where she subsequently carried out her field research, knows first-hand about the gender dynamics described above by Scott (Smith Rolston 2010). Unlike the male miners of Carrara, female miners in Wyoming are deeply dissatisfied with their work. This dissatisfaction is linked to the fact that their knowledge and skills are often underestimated by colleagues, especially men, who assume that women necessarily have little affinity for technology. Thus, when they are uncertain or make mistakes, they confirm this underlying stereotype, while when they prove to be competent, they are accused of being bossy. This dynamic also emerges in the radio conversations between miners – conversations that Smith Rolston examines through queer theory applied to linguistics. Her approach aims to show, on the one hand, the complex process of co-production of gender and technology and, on the other, how this same process cannot be reduced to a binary relationship between men and women alone. Through a detailed ethnographic analysis of miners' discourses, Smith Rolston shows, if anything, how "in one of the allegedly most heavily masculinised industries, women have cultivated alternative ways of being feminine" (ibid.: 914).

Emphasizing the skills that miners demonstrate when handling their tools, and the pride – often, in a hypermasculine form – that comes from the awareness of these same skills, should not make us forget how frustrating or physically suffocating and overwhelming mining can be. In a very different context from that of the Carrara marble quarries and the coal mines of West Virginia or Wyoming, namely in the Bolivian Andean mountains, the mountain is not an enemy to be beaten or to be flattened, but an entity to be feared. Just as men *eat* the mountain, the mountain itself *eats* men. In *We Eat the Mines and the Mines Eat Us* ([1979]1993), June Nash reports the words of an expert tin miner who confided to the anthropologist:

> After seventeen years of working, when I get near the mine, I can still smell the sulphates and I feel suffocated by the gases (...). Although I haven't used the drill for two years, I can still hear the sound of the air hissing through the hoses when I lie down to go to sleep.
>
> *(Nash [1979]1993: 194–195)*

These words reveal how technology can be literally incorporated to the point of becoming a persecutory presence that metaphorically eats bodies rather than an extension of the self to be proud of. South African miners' ears know something about loud and repetitive noise made by drills and other machines too. Years of underground work in the gold mines make them gradually lose the capacity to differentiate sounds and eventually lead to deafness (Morris 2008). Juxtaposed with marble quarrymen, the words of Bolivian miners also tell us something about the fact that the self-representation of miners, and their relationship with technology, are conditioned not only by their different social positions, but also by the cultural and economic context in which they live as well as by the ideological background through which they comment on their profession (cf. Scott 2010).

Drillers who work underground for large-scale mining companies are also dealt with by the historian Dunbar Moodie in collaboration with Vivienne Ndatshe. Among the many questions contained in *Going for Gold* (1994), they examine the reaction of South African miners to the introduction of new drilling technologies in the early nineteenth century. This study is relevant here because it allows us to consider another important aspect that is often underestimated in anthropological analyses of extractive industries, namely, the question of the relationship between capital, technology, and race. For a long time, in the South African mines, tasks were divided not only on the basis of the specific skills and experiences of each worker, but also on the basis of skin colour or ethnicity (see also Guy and Thabane 1988). Until the 1980s, drilling operations were reserved for black labour, while the use of explosives was permitted, by law, exclusively to white miners.

Moodie and Ndatshe's attention is focused on the years following the end of the First World War when manual drilling was considered by companies' managers to be safer and more convenient than drilling with machines. In fact, hammer drillers were not only more accurate and efficient, but they also required less supervision than machine drillers. Supervisors were all white men, they were paid much better, and for this reason, they were more expensive than black labour. Unsurprisingly, white unions were among the main supporters of the mechanization of mining operations. Thus, the decision to continue to use hammer drillers was motivated not by a supposedly better efficiency of mechanized labour – which was the main argument of the white unions – but by the lack of black labour willing to accept the wages offered by the company for manual drilling. However, machine drillers produced greater quantities of rubble and therefore required more people to clear the workspaces. In the absence of larger lashing teams, the managers asked the drillers to take care of these clearing operations themselves. This request caused discontent among the drillers. They were paid on a contract basis and this job, which was not part of their duties, was unpaid. Thus, it was also necessary to gradually mechanize the work of clearing the rubble. In short, the introduction of machine drillers was not an immediate and obvious process. It took a few decades for their use to become obvious. Moreover, it involved complex negotiations between managers, black and white workers, as well as gradual organizational and technical adjustments.

To conclude, the cases examined in this section remind us that the bodies of miners are privileged terrains in which, and through which, relationships of power and domination are inscribed. The (hyper-toxic-) masculine self-representations of Italian marble quarrymen or American coal miners counterbalance a feminine nature to be tamed with the masculine power of technology. This brings us back to the question of the relationship between technology and gender. In this regard it is interesting to note that, numerically, the presence of women in mines is, in many cases, comparable to that of men. Still, there is a glaring tendency on the part of social scientists to focus on male presence – a trend that various scholars have begun to question (e.g. Lahiri-Dutt 2015) as some of the studies mentioned here also testify. Moreover, South African gold miners remind us that their bodies are not only sexualized, but also racialized. This social dimension would also deserve further study as would, more generally, anthropological studies on mining technology that intersect questions of gender, race, class, and any other significant axis of analysis relevant in each case.

5 Conclusion

This chapter has highlighted how mining technology is the result of creative combinations of heterogeneous, technical, and non-technical elements, which often have a high degree of mobility. Knowledge, experiences, tools, people, and ideas travel across time and space, and, as they travel, some can be partly forgotten, perhaps to be rediscovered later. Often, they combine with other technologies and experiences in ways that are not always predictable. The ethnographic perspective promoted here enters into a dialogue with different disciplines to show how these technological combinations arise, as well as how mining technologies contribute to shaping not only specific environments, but also miners' subjectivities and relations of class, gender, and race. In short, it illustrates in the most detailed possible way, how mining technology is the result of multiple assemblages that contribute to generate and modify specific forms of life in the mines and around them (cf. Pfaffenberger 1988; Winner 1986).

As anticipated at the beginning of this chapter, there are still many other issues waiting to be explored or deepened with regard to the relationship between mining and technology. A central question remains that of understanding the role of mining technologies in assisting the processes of extension and intensification of mining capitalism. By extensive processes we mean here those processes that aim to establish and pursue ever new extractive frontiers, however fictitious they may be. On the other hand, by intensification processes, we mean those socio-technological processes that contribute to the maximum possible exploitation of the mineral potential of a specific environment. Both processes are closely intertwined and confront the problem of the limit to environmental exploitation, but they differ in the type of response they offer. Extensive processes respond to this problem by denying it or moving it towards a new extractive frontier. Deep seabed mining and asteroid or space mining are recent examples of this type of response. Until a few years ago, asteroid mining appeared to be pure science fiction. Yet, the multiplication of workshops, conferences, and legal and commercial initiatives aimed at securing the mineral rights and technologies necessary for future space explorations on asteroids and Earth's Moon signal how this type of "extraglobal" extraction is less unlikely than it was until recently (e.g. Klinger 2017).

Deep seabed mining, on the other hand, is already operational thanks to the recent development of very expensive remote-manoeuvrable machinery. However, technological advances do not seem to have gone hand in hand with the development of an adequate regulatory regime for mineral exploration (e.g. Le Meur et al. 2018). Some scholars have already highlighted the potential devastating effects of underwater mining technologies on marine ecosystems, as well as for human communities living along the coasts most directly affected (e.g. Carvera et al. 2020). In many of these studies, however, the technology itself remains in the shade – partly due to the reluctance of mining companies to reveal important details about how these machines work (see Chin and Hari 2020).

As for the intensive side of mining, the historic attempt to make mining technologies increasingly efficient and productive, and at the same time delink them from human labour, has recently been matched with the need to make them "greener". Beyond techno-optimistic narratives, one wonders, however, to what extent human work can be automated in the mines. In some large-scale mines it is already possible to observe machinery or trucks being operated from offices located long distances away (e.g. Paredes and Fleming-Munoz 2021), but it remains to be seen how far these automation processes can go. When it comes to the extensive processes of automation and robotization of the extractive industry, it does not seem that the various levels of this same industry are taken into account. Will we ever see robots in the alluvial mineral deposits around the world from which millions of miners currently use their hands and artisanal tools to earn an income? There is a tendency to give a lot of emphasis to processes of change which, in reality, touch a specific extractive level. Instead, as suggested here, it would be more profitable to consider the mining industry as a whole and examine technological changes in terms of additions rather than substitutions. From this perspective, an issue that arises is to understand the reasons for the coexistence or the articulation in the mining industry of the wooden sieve used by artisanal miners, and the hyper-technological probe which is controlled thousands of miles away with the support of artificial intelligence.

Another area worthy of anthropological investigation is that of biotechnologies applied to the exploration, extraction, and stabilization of mining waste. These are largely experimental technologies that seek to find sustainable solutions for mining: microorganisms that reveal the existence of mineral deposits, or that can be used to reduce the toxicity levels of waste, as well as plants cultivated to assimilate the precious minerals left in the residues of mining companies (Wilson-Corral et al. 2012).[3]

In general, it would be desirable that all these fields dominated by geological, geochemical, and engineering studies were accompanied by anthropological analyses – even if only to examine the futuristic or techno-optimistic narratives surrounding mining technology. In this regard, it is worth highlighting an assumption that seems to be common to recent developments in mining technologies: extractive activities create various types of problem; however, these can be solved by advances in technology (LeCain 2004). As Rajak (2020) notes, the recent wave of "techno-optimism" seems to have so far translated into a sort of magical thought in which technology is a *deus ex machina*. Indeed, it appears capable of reconciling, on the one hand, the uncontainable capitalist desire to grow indefinitely and, on the other, the need to preserve, at least in part, the material conditions of this same growth, that is the environment. However, this attitude risks leaving things as they are or even making them worse without critically examining the complex relationships between society, resource extraction, and technology.

Acknowledgements

I would like to thank Eleanor Fisher, Robert J. Pijpers, and Benjamin Rubbers for commenting on earlier versions of this chapter. Errors and inaccuracies are my only responsibility. The writing of this chapter has in part benefitted from a contract at the University of Reading, UK between 2019 and 2020 (Gold Matters [ref: 462.17.201] financed by the Belmont Forum and Norface Joint Research Programme on Transformations to Sustainability, which is co-funded by DLR/BMBF, ESRC, FAPESP, ISC, NWO, VR, and the European Commission through Horizon 2020).

Notes

1 This definition of mining technology as assemblage inspired the panel that Michael Bürge and I organized at the EASA 2014 conference entitled "Mining technology: practices, knowledge and materials across and beyond the mines". I am grateful to the participants who made this panel possible with their contributions.
2 As Nash ([1979]1993) notes this is particularly evident in large-scale mining where female labour is most often affected.
3 The biotechnology literature is expanding rapidly, and dedicated research groups have been formed, such as the Mining Microbiome Research Group of the University of British Columbia.

Bibliography

Akiwumi F. and D'Angelo L. (2018) The Sierra Leone rare earth minerals landscape: An old or new frontier? *The Extractive Industries and Society* 5 (1): 36–43.
Bijker W.E., Hughes T.P., Pinch T. and Douglas D.G. (eds) 2012) *The social construction of technological systems: New directions in the sociology and history of technology*, Cambridge: MIT Press.
Bray F. (2007) Gender and technology. *Annual Review of Anthropology* 36: 37–53.
Calkins S. (2016) How "clean gold" came to matter: Metal detectors, infrastructure, and valuation. *HAU: Journal of Ethnographic Theory* 6 (2): 173–195.
Carvera R., Childs J., Steinberg P., Mabonc L., Matsuda H., Squire R., McLellan B. and Esteban M. (2020) A critical social perspective on deep sea mining: Lessons from the emergent industry in Japan. *Ocean & Coastal Management* 193 (1): 105242.
Chin A. and Hari K. (2020) *Predicting the impacts of mining of deep sea polymetallic nodules in the Pacific Ocean: A review of the scientific literature*, Deep Sea Mining Campaign and Mining Watch Canada.
Curtin P. (1973) The lure of Bambuk gold. *The Journal of African History* 14 (4): 623–631.
D'Angelo L. (2018) Diamonds and plural temporalities: Articulating encounters in the mines of Sierra Leone. In Pijpers R.J. and Eriksen T.H. (eds) *Mining encounters. Extractive industries in an overheated world*, London: Pluto Press, 138–155.

D'Angelo L. (2019) *Diamanti. Pratiche e stereotipi dell'estrazione mineraria in Sierra Leone*, Milan: Meltemi [Diamonds. Stereotypes and Mining Practices in Sierra Leone].

d'Avignon R. (2018) Primitive techniques: From "customary" to "artisanal" mining in French West Africa. *Journal of African History* 59 (2): 179–197.

De Rijke K. (2018) Drilling down comparatively: Resource histories, subterranean unconventional gas and diverging social responses in two Australian regions. In Pijpers R.J. and Eriksen T.H. (eds) *Mining encounters: Extractive industries in an overheated world*, London: Pluto Press, 97–120.

Dessertine A.(2016)From pickaxes to metal detectors: Gold mining mobility and space in Upper Guinea, Guinea Conakry. *The Extractive Industries and Society* 3(2): 435–441.

Dumett R.E. (1998) *El Dorado in West Africa: The gold-mining frontier, African labor, and colonial capitalism in Gold Coast, 1875–1900*, Oxford: James Currey.

Fisher E. (2008) Artisanal gold mining at the margins of mineral resource governance: A case from Tanzania. *Development Southern Africa* 25 (2): 199–213.

Godoy R. (1985a) Mining: Anthropological perspectives. *Annual Review of Anthropology* 14: 199–217.

Godoy R. (1985b) Technical and economic efficiency of peasant miners in Bolivia. *Economic Development and Cultural Change* 34 (1): 103–120.

Golub A. (2019) Mining. In: Stein F., Lazar S., Candea H., Diemberger J., Robbins A., Sanchez A. and Stach R. (eds) *The Cambridge encyclopedia of anthropology*, http://doi.org/10.29164/19mining.

Greenhalgh P. (1985) *West Africa diamonds 1919–83: An economic history*, Manchester: Manchester University Press.

Guy J. and Thabane M. (1988) Technology, ethnicity and ideology: Basotho miners and shaft-sinking on the South African gold mines. *Journal of Southern African Studies* 14 (2): 257–278.

Hilson G. and Pardie S. (2006) Mercury: An agent of poverty in Ghana's small-scale gold mining sector? *Resource Policy* 31: 106–116.

Hilson G. and Van der Vorst R. (2002) Technology, managerial, and policy initiatives for improving environmental performance in small-scale gold mining industry. *Environmental Management* 30 (6): 764–777.

Hornborg A. (2013) Symbolic technologies: Machines and the Marxian notion of fetishism. *Anthropological Theory* 1 (4): 473–496.

Ingold T. (2000) *The perception of the environment: Essays on livelihood, dwelling and skill*, LondonandNew York:Routledge.

Ingold T. (2011) *Being alive: Essays on movement, knowledge and description*, London – New York: Routledge.

Isenberg A.C. (2005) *Mining California: An ecological history*, New York: Hill and Wang Press.

Kama K. (2020) Resource-making controversies: Knowledge, anticipatory politics and economization of unconventional fossil fuels. *Progress in Human Geography* 44 (2): 333–356.

Kasuga N. (2010) Total social fact: Structuring, partially connecting, and reassembling. *Revue du MAUSS* 36 (2): 101–110.

Killick D. and Fenn T. (2012) Archaeometallurgy: The study of preindustrial mining and metallurgy. *Annual Review of Anthropology* 41: 559–575.

Klinger J.M. (2017) *Rare earth frontiers: From terrestrial subsoils to lunar landscapes*, Ithaca and London: Cornell University Press.

Knapp A.B., Pigott V.C. and Herbert E.W. (eds) (2002) *Social approaches to an industrial past. The archeology and anthropology of mining*, London and New York: Routledge.

Lahiri-Dutt K. (2015) The feminisation of mining. *Geography Compass* 9 (9): 523–541.

Lanzano C. (2020) Guinea Conakry and Burkina Faso: Innovations at the periphery. In: Verbrugge B. and Geenen S. (eds) *Global gold production touching ground*, Cham: Palgrave Macmillan, 245–262.

Latour B. (2005) *Reassembling the social. An introduction to Actor-Network-Theory*, Oxford: Oxford University Press.

LeCain T.J. (2004) When everybody wins does the environment lose? The environmental techno-fix in twentieth century American mining. In: Rosner L. (ed.) *The technological fix: How people use technology to create and solve problems*, New York: Routledge, 137–153.

Leitch A. (1996) The life of marble: The experience and meaning of work in the marble quarries of Carrara. *The Australian Journal of Anthropology* 7 (3): 235–257.

Le Meur P-Y., Arndt N., Christmann P. and Geronimi V. (2018) Deep-sea mining prospects in French Polynesia: Governance and the politics of time. *Marine Policy* 95: 380–387.

Lemonnier P. (1996) Technology. In: Barnard A. and Spencer J. (eds) *The Routledge encyclopedia of social and cultural anthropology, II edition*, London and New York: Routledge, 684–688.

Luning S. (2014) The future of artisanal miners from a large-scale perspective: From valued pathfinders to disposable illegals. *Futures* 62A: 67–74.

Luning S. (2018) Mining temporalities: Future perspectives. *The Extractive Industries and Society* 5 (2): 281–286.

Luning S. and Pijpers R.J. (2017) Governing access to gold in Ghana: In-depth geopolitics on mining concessions. *Africa* 87 (4): 758–779.

Maia F., Veiga M.M., Stocklin-Weinberg R. and Marshall B.G. (2019) The need for technological improvements in Indonesia's artisanal cassiterite mining sector. *The Extractive Industry and Society* 6 (4): 1292–1301.

Manuel J.T. (2017) Efficiency, economics, and environmentalism: Low-grade iron ore mining in the Lake Superior District, 1913–2010. In: McNeill J.R. and Vrtis G. (eds) *Mining North America: An environmental history since 1522*, Oakland: University of California Press, 191–216.

Maponga O. and Ruzive B. (2002) Tribute chromite mining and environmental management on the Northern Great Dyke of Zimbabwe. *Natural Resources Forum* 26: 113–126.

Massaro L. and de Theije M. (2018) Understanding small-scale gold mining practices: An anthropological study on technological innovation in the Vale do Rico Peixoto (Mato Grosso, Brazil). *Journal of Cleaner Production* 204: 618–635.

Moodie T.D. and Ndatshe V. (1994) *Going for gold. Men, mines, and migration*, Berkeley, Los Angeles and London: University of California Press.

Morris R.C. (2008) The miner's ear. *Transition* 98: 96–115.

Naji M. and Douny L. (2009) Editorial. *Journal of Material Culture* 14 (4): 411–432.

Nash J. ([1979]1993) *We eat the mines and the mines eat us. Dependency and exploitation in Bolivian tin mines*, New York: Columbia University Press.

Ofosu-Mensah E.A. (2017) Mining in Ghana and its connections with mining in the Brazilian diaspora. *The Extractive Industries and Society* 4 (3): 473–480.

Paredes D. and Fleming-Munoz D. (2021) Automation and robotics in mining: Jobs, income and inequality implications. *The Extractive Industries and Society* 8 (1): 189–193.

Pfaffenberger B. (1988) Fetishised objects and humanised nature: Towards an anthropology of technology. *Man* 23: 236–252.

Pfaffenberger B. (1992) Social anthropology of technology. *Annual Review of Anthropology* 21: 491–516.

Pfaffenberger B. (2002) Mining communities, chaînes opératoires and sociotechnical systems. In: Knapp A.B., Pigott V.C. and Herbert E.W. (eds) *Social approaches to an industrial past. The archaeology and anthropology of mining*, London and New York: Routledge, 291–300.

Pijpers R. (2020) Ghana: A history of expansion and contraction. In: Verbrugge B. and Geenen S. (eds) *Global gold production touching ground*, Cham: Palgrave Macmillan, 185–206.

Rajak D. (2020) Waiting for a *deus ex machina*: "Sustainable extractives" in a 2°C world. *Critique of Anthropology* 40 (4): 471–489.

Ross C. (2014) The tin frontier: Mining, empire, and environment in Southeast Asia, 1870s–1930s. *Environmental History* 19: 454–479.

Russell-Wood A.J.R. (1977) Technology and society: The impact of gold mining on the institution of slavery in Portuguese America. *The Journal of Economic History* 37 (1): 59–83.

Saugeres L. (2002) Of tractors and men: Masculinity, technology and power in a French farming community. *Sociologia Ruralis* 42 (2): 143–159.

Scharff R.C. and Dusek V. (eds) (2003) *Philosophy of technology. The technological condition. An anthology*, Malden, MA and Oxford: Blackwell.

Scott R.R. (2010) *Removing mountains. Extracting nature and identity in the Appalachian coal-fields*, Minnesota: University of Minnesota Press.

Seccatore J. and de Theije M. (2017) Socio-technical study of small-scale gold mining in Suriname. *Journal of Cleaner Production* 144: 107–119.

Silver J. (1981) The failure of European mining companies in the nineteenth-century Gold Coast. *The Journal of African History* 22: 511–529.

Sinding K. (2005) The dynamics of artisanal and small-scale mining reform. *Natural Resource Forum* 29: 243–252.

Smith Rolston J. (2010) Talk about technology: Negotiating gender division in Wyoming coal mines. *Sign* 35 (4): 893–918.

Teschner B., Smith N.M., Borrillo-Hutter T., Quaghe John Z. and Wong T.E. (2017) How efficient are they really? A simple testing method of small-scale gold miners' gravity separation systems. *Minerals Engineering* 105: 44–51.

Thabane M. (2003) Technical skills, technology and innovation among individual diamond diggers at the Letšeng-la-Terae and Kao diamond mines of Lesotho, 1960–1970. *South African Historical Journal* 49: 147–161.

Thoburn J. (1977) Commodity prices and appropriate technology: Some lessons from tin mining. *The Journal of Development Studies* 14 (1): 35–52.

Veiga M.M., Angeloci-Santos G. and Meech J.A. (2014) Review of barriers to reduce mercury use in artisanal gold mining. *The Extractive Industries and Society* 1 (2): 351–361.

Verbrugge B. (2020) Technological innovation and structural change. In: Verbrugge B. and Geenen S. (eds) *Global gold production touching ground*, Cham: Palgrave Macmillan, 97–114.

Verbrugge H. and Van Wolputte S. (2015) Just picking up stones: Gender and technology in a small-scale gold mining site. In: Coles A., Gray L. and Momsen J. (eds) *The Routledge handbook of gender and development*, Abingdon and New York: Routledge, 173–184.

Warnier J-P. (2009) Technology as efficacious action on objects … and subjects. *Journal of Material Culture* 14 (4): 459–470.

Werthmann K. (2010) Following the hills: Gold mining camps as heterotopias. In Freitag U. and von Oppen A. (eds) *Translocality. The study of globalising processes from a southern perspective*, Brill: Leiden, 111–132.

Werthmann K. and Grätz T. (eds) (2012) *Mining frontiers in Africa. Anthropological and historical perspectives*, Cologne: Rüdiger Köppe Verlag.

Wilson-Corral V., Anderson C.W.N and Rodriguez-Lopez M. (2012) Gold phytomining: A review of the relevance of this technology to mineral extraction in the 21st century. *Journal of Environmental Management* 111: 249–257.

Winner L. (1986) *The whale and the reactor: A search for limits in an age of high technology*, Chicago: University of Chicago Press.

11

UNDERGROUND

Sabine Luning

1 Introduction: the politics of the vertical turn

In recent years, scholars from a range of disciplines have begun to emphasize that analysis of political forms of domination, the effects of technological innovations, and economic developments in resource extraction requires a vertical view. This vertical turn has meant that, rather than merely examining surface fields, scholars are now taking into account three-dimensional space: moving from a focus on area to one on volume (Elden 2013) and from studies on the politics of mapping to cartographies that capture the politics of verticalism (Weizman 2002). Whether in geography, STS, urban studies, or anthropology, the volumes of the subterranean, the seas, and the air are a growing focus, as well as how they feature in knowledge systems, governance practices, and technologies of access (Battaglia 2020; Billé 2017, 2019).

In long-term, global, and often colonial histories the underground has been targeted for extracting minerals, but it also hosts vertical urban developments (Harris 2015) as well as storage of waste and placement of infrastructure (Bélanger 2007; Chilvers and Burgess 2008; Morris 2008a). With deep sea mining, resource extraction has recently become much more common under extraterritorial waters, where the building of oil platforms and the construction of infrastructure for communication systems (cables) and wind energy lead to what can be called an urbanization of the sea (Couling and Hein 2020). The air, on the other hand, is the zone of traffic, surveillance technologies, and the distribution of property rights for communication systems (Adey 2015). This proliferation of vertical developments and the securing of volumes at different heights and depths has led to a rise in studies presenting interesting connections between disciplinary fields and between underground and above ground space. In urban studies, attention to the vertical dimensions of building projects (Graham 2016) is combined with tracing the dependence of cityscapes on underground extraction. Furthermore, modern cities

DOI: 10.4324/9781003018018-11

can be conceptualized as concrete urban jungles built on the basis of bauxite, iron, and plastics or even described as the "inverted mines" (Brechin 2006) of distant resource hinterlands. This latter idea has been taken up and politicized by Elizabeth Povinelli (2019), who indicates how oppressive extractive practices elsewhere can literally be brought home to city-dwellers.

Political analysis is a key element in many studies concerned with the race to the bottom or that into the sky, since these trends epitomize capitalist growth drives, with massive implications for the futures of contemporary societies, notably in terms of climate change, social justice, and sustainable development. Discussions of the pros and cons of the term Anthropocene highlight the political dimensions of the vertical turn as its definition emphasizes the vertical nexus between geology and society, while the more specific term "Subterranean Anthropocene" has been coined to put the underground centre stage in the debates (Melo Zurita, Munro, and Houston 2018). The latter term foregrounds how the climate-change wake-up call is shaped by imaginaries of the underground in terms of geological time-scales established through stratigraphy – the study of sedimentary layers (e.g. rock, ice) (ibid. p. 2). The nexus between geology and society, implicit in the notion of the Anthropocene but also more generally in ways of thinking about human and non-human life on earth, is currently a major issue for critical political reflection. What characterizes the distinction between the geological and the social, the human and the non-human? And how are these categories deployed in timeframes for thinking about destruction of the earth, of life, and for identifying systems and actors responsible for it and those suffering from it? Alternative terms such as Capitalo-cene (Moore 2017) and Black Anthropocenes (Yusoff 2018) radicalize the wake-up call of climate change by focusing on modes of exploitation of the earth and of specific groups of people in longer histories of colonial and racialized practices of extraction. Indeed, extraction from the underground has long been implicated in processes of world making and world breaking: wealth creation, projects of urba-nization, and technological innovations, as well as counterpoints in the suffering of miners, as slaves and labourers, and the planet itself (Sheller 2014; Yusoff 2015).

The political debates interrogating the nexus between geology and the social draw attention to vertical social hierarchies as well as the vertical spatial dimensions of extraction. These upsides and downsides of mining have been discussed exten-sively in the literature on the curses and blessings of resource extraction (Ross 1999), but these aspects can be refined and radicalized from in-depth perspectives, both spatial and historical. Extracting from the underground goes hand in hand with ways of knowing which implicate morals: what characterizes the world of the underground and how can people take from it without risking their lives? How can the downsides of large-scale mining be legitimized by its upsides? Or are the costs always too high – for people and the environment – and should we just keep potential mineral wealth in the ground (Khatua and Stanley 2006)? Furthermore, who is "we" in the latter option? Humanity at large, no doubt, but who is inclu-ded and excluded by this term? Delving into the underground is accompanied by social and cultural diversities in knowledge, power, and positioning. As the discipline which puts cultural dimensions and social diversity centre stage, anthropology has specific contributions to make in this vertical terrain of the social and subterranean, given that its hallmark is the analysis of plurality in ways of knowing and how these are socially embedded. This attention is reflected in the anthropology of the underground, primarily in connection with questions of diversity in ontological and cosmological perspectives, and perceptions of the underground as a source of potential wealth as well as a site of danger and suffering.

In this chapter, I first engage with these more general questions of ways of knowing the underground and moral issues connected to them. How to know and value the underground and its potential is important. Therefore, special attention is given to the senses in which people can know this subterranean world; the capacity to see or not to see mineral wealth features prominently in discourses on the underground as a subject of knowledge. Moreover, even though the chapter is primarily concerned with the underground, water is also important in a vertical analysis of extraction, both as resource located underground and as subsurface volume for deep sea mining. After the general overview, the chapter moves on to examine innovative ethnographic work in which the underground takes centre stage. The latter will focus on three themes: governance, mining companies, and work practices in – mainly large-scale – extraction of minerals.

2 Cosmologies: underground mining as relations of exchange

Anthropology which focuses on mining the underground has followed a trajectory which also characterizes other fields, such as the study of agriculture, and disciplines: moving from a focus on labour to land. Since the 1930s the work by members of the Rhodes-Livingstone Institute on mining and urbanization in Southern Africa has changed anthropology drastically: social change started to be studied from the angle of fast industrialization, as a movement from rural to urban, from tribesman to townsman, and from farmer to miner (Mayer 1961; Mitchell 1959). The effects on social formations of new forms of labour – both in their oppressive colonial practices and their new opportunities for social bonding – changed the discipline of anthropology far beyond the field of mining (Ferguson 1999). Despite work practices underground featuring prominently in the latter, the underground as a peculiar space of knowing and working was hardly ever central.

The first time the ways the underground was perceived, known, and targeted by miners was studied ethnographically was in June Nash's now classic work on Bolivian tin mines, significantly entitled *We Eat the Mines and the Mines Eat Us* (Nash 1979). The title expresses how miners are entangled in relations of agency with the mines that are marked by giving and taking. Describing one of the many rituals carried out by mining communities both above and below ground, Nash explains the title of the book as follows:

> At level 340 the men gathered around the llamas as the *yatiri* cast liquor in all
> directions, asking the *awichas*, the *malcus*, the *tíos* to protect each level in the

mine, beginning with level 340, pleading with the *Tío* not to eat any more workers and that he yields them much mineral.

(ibid.: 157)

The mines are considered spaces of danger as well as potential wealth, and in order to avert risk and obtain yields relations of exchange with spirits of the mines are in order. Nash pays substantial attention to work practices – she describes most vividly how she herself went underground for whole working shifts – and connects this with how the miners perceive these underground spaces. In this regard she notes that it is not only the spirits that imbue the underground world with agency; the caverns of the mine are also seen as living organisms and the act of extraction is equated with inflicting wounds to drain the fluids of mineral wealth (ibid.: 179; see also Ferry this volume). Acts of exchange are important in these working spaces marked by danger, tensions, and inflicted violence. June Nash analyses these cosmologies of the underground as part of contemporary global capitalisms which have been shaped and transmitted in longer histories of colonial extraction. She states that the exchanges with underground powers should be seen as contemporary manifestations of precolonial moralities (ibid.: 27) and compared to commodified capitalist ways of taking from the underground. In her ethnography of labouring underground she foregrounds the dangers, the inequalities, and the suffering to which miners are subjected.

Michael Taussig (1980) uses the work of June Nash to further the comparison between the way Bolivian tin miners engage in rites of sacrifice and gift exchange, and commodified capitalist models of exchange. He demonstrates that commodification requires categories of space, time, causation, and human relations to be presented as "natural things", rather than as embedded in historical and social contexts (D'Angelo 2016). He also describes historical shifts in valuing underground mining terrains, particularly the shift from cosmologies in which the underground is associated with agency and gift exchange, to those that posit the underground as dead matter which can be priced as a commodity. Taussig's work, which explicitly takes lessons (1980: 8) from Polanyi's *The Great Transformation* (1944), inspires current anthropological thinking about ontologies and the critique of resources as "natural things". Polanyi shows that labour and land do not in and of themselves exist in commodified forms; rather, it requires considerable efforts by state legislation and otherwise to transform them into market-ready commodities.

This line of analysis has been taken up in interesting ways in the conceptualization of "mineral resources" in underground scholarship where resources are often talked about in naturalized ways, as discrete entities in nature waiting to be exploited. Anthropologically, however, it is important to consider subterranean territory that is targeted for mineral extraction as "socio-nature" (Gellert 2005), as orebodies embedded in specific social and technical frames of possibilities and circumstances. Richardson and Weszkalnys (2014) also demonstrate that resources should be studied as entanglements of the material with the social, as the way in which professional miners distinguish "resources" from "reserves" illustrates how chunks of earth are commodified as socio-natures. Whereas a "resource" and a "reserve" are both considered potentially valuable deposits of mineralized ore, extraction of a "reserve" is considered feasible given the actual legal, political, technical, and economic circumstances in the present. Changes in jurisdiction, political circumstances, technical developments, and the world market price of a specific mineral can all play a role in re-classifying "ore" as either reserve or resource, and occasionally even stop projects completely (Luning and Pijpers 2017; Sosa and Zwarteveen 2012). These anthropological contributions are an in-depth extension of Polanyi's concerns regarding embedding and disembedding chunks of human energy and land. Making chunks of earth into commodified resources depends on a double movement of disembedding mining matter from the wider landscape by embedding it in social circumstances.

This overview of the trajectory from Polanyi through Taussig to the current attention to "making resources" allows us to frame different ways of knowing and engagements with the underground as varieties of cosmological relations of exchange.

3 Underground ontologies: (in)visibility and the politics of classification

Promising anthropological lines of research emerge from the analytical frame which considers cosmologies of the underground as varieties of exchange relations. The frame foregrounds that different ways of knowing – often positioned oppositionally as scientific versus cultural – are underpinned by ontological postulates about life and non-life, and the agency or inertness of underground matter. These inform, first of all, considerations concerning the capabilities for identifying and appropriating underground wealth. Secondly, as we will see when discussing work of Povinelli and Yusoff, ontological postulates have political dimensions which are constitutive of socio-political verticalities between knowledge systems and within histories of mining practices.

Among the senses with which people are capable of identifying underground mineral wealth, sight is given a special position. Cosmologies which attribute value to the underground by embedding chunks of dead matter in social circumstances stress that mineral wealth is invisible to humans on the ground. Many studies – as we will see in the course of this chapter – highlight invisibility as a starting point for "making resources". Whether underground minerals or underground water (aquifers), humans on the surface cannot see subsurface. Based on fieldwork in Costa Rica, Andrea Ballestero (2019) analyses how aquifers become a major source of water supply by embedding its distribution in legal and technical structures. How water should be valued – as public good and private property – depends on knowledge about the quality and quantity of this underground resource; however, invisibility and form (aquifers are liquid and in movement) make this hard to determine. It is difficult to imagine an aquifer as a figure that can be separated from the ground. Since aquifers cannot be seen, people depend on conceptualizations, calculations, and modelling which allow the assessment of the aquifer's characteristics, because envisioning aquifers and knowing how to tap into them are important for public debates and reliable decision making concerned with water sources.

In cosmologies which, on the other hand, attribute underground wealth with agency, appropriation of mineral wealth depends on the ability "to see it". When dealing with tensions between mining teams competing over ore, June Nash describes how witchcraft may be used to prevent competitors from seeing the ore, or to make the ore disappear (Nash 1979: 160). That underground wealth has to allow itself to be seen is a recurring theme in the ethnography of gold mining in West Africa. Gold is a "substance" (van de Camp 2016), which should be approached and handled with care (Werthmann 2003); it has living qualities and it can move, it must allow itself to be found/seen, and the spiritual owners (spirits of the bush and ancestors) have to consent to its appropriation (Luning 2012). Miners often consult so-called "bush-doctors", diviners who have double sight – "four" eyes – allowing them to see features of the world of the bush, bush-beings, and the underground which are invisible to others (Bonnet 1988). Lorenzo D'Angelo, writing about diamond mining in rivers in Sierra Leone, describes the dangerous work of artisanal miners who dive and dig in the darkness of the riverbed with their eyes closed. They see by touching and construct a mental map of the environment as a support to prevent risks, but even the most skilled divers are aware of the visible and invisible threats. Invisible spiritual beings (djinns) rank among the dangers, capricious beings that could grant but also withhold their riches, regardless of the offerings made by the miners (D'Angelo 2019: 52). D'Angelo also discusses the importance of relations between the senses – sight and touch in this case – a subject to which we return later.

For now, these examples show the prominent place of the distinction between visibility and invisibility, between sight and the capacity to take from underground and underwater sources of wealth, and between cosmologies foregrounding, respectively, underground agency or inert, non-life matter. In current studies, hybrid situations in which these cosmological perspectives occur simultaneously are analysed to see how they may be magnified, hierarchized, delimited, and co-constituted. Many studies demonstrate that cosmological ideas which allocate agency to underground wealth and beings can be put to work and amplified in situations of contestation between local communities (e.g. of small-scale miners) and large-scale mines. Case studies are situated in Ghana (Rosen 2020) and Latin America (De la Cadena 2015), but ethnography on Papua New Guinea also testifies to these processes, with the Porgera mine providing an emblematic site of contestations (Golub 2014; Jacka 2015). Importantly, in many mining encounters (cf. Pijpers and Eriksen 2018) the cosmological perspectives of local communities and mining companies bring into play different assumptions about distinctions between human/nonhuman, life/non-life in underground worlds. In her recent work, Povinelli (2016) proposes the concept "geontologies" to show how, in Australia, colonial settlers' categorization of extractive landscapes as Nonlife has served to ignore and belittle the different categorizations of Life and Nonlife applied in Aboriginal societies.

Povinelli's work shows the political dimensions of cosmological perspectives and their distinctions between life and dead matter. The politics of classifications is also a recurring theme in debates on the Anthropocene. Pertinent to knowledges of the underground is the analytical contribution by Catherine Yusoff, who holds a chair in "Inhuman Geography" at Queen Mary University of London. Yusoff takes the histories of commodification of labour and land in colonial geological endeavours as a starting point for a radical critique of the term Anthropocene. Highlighting the racialized histories in Latin America, in which slavery served as a major institution in gold mining projects, she criticizes assumptions about the nexus between geology and society, and between Life and Nonlife, as well as the historical timelines deployed in debates on the Anthropocene. In her view geology is not an innocent or natural description of the world, but "a discipline and extraction process cooked together in the crucible of slavery and colonialism" one historically situated as a transactional zone (Yusoff 2018: Chapter 1, Geology, Race and Matter). This formulation redefines geology as the socio-nature of exchange and suffering: labouring bodies have been and still are sacrificed in mines (Yusoff 2015) and geology is not merely concerned with pre-economic dead matter, but a practice in which wealth is taken at the expense of the "environment", but also of groups of "inhumanized" people. For these subjects, the end of this world has already happened. She sees the Anthropocene as an origin story for white liberal communities' exposures to environmental harm. This attempt to decolonize the study of underground extraction stretches the meaning of the term extraction by applying it to historical relations between people (See also Gómez-Barris 2017), meanwhile bringing ways of knowing and ways of taking into a single frame. Moreover, it does so with a keen eye for both commodified labour and land.

Perhaps the Rhodes-Livingstone Institute may have over-emphasized the labour side of underground mining, but the work of Yusoff warns us that currently some anthropologists may have started to succumb to tunnel vision on underground and environmental matters. Retaining exchange modalities as a historical trajectory of attention since Polanyi's work can help the discipline to keep a balanced view on labour and land as part of dynamic, plural, and power-loaded vertical socio-natures. Since Polanyi, anthropologists have also been taking stock of the fact that making resources takes a lot of effort; governing authorities need to put in place rights and regulations for underground wealth to be tackled by miners working on different scales. The next section is devoted to governing the underground, as an important (legal) practice in the making of resources.

4 Governing the underground

Interest in vertical dimensions of governing practices has recently been given interdisciplinary attention in books with titles such as *Voluminous States* (Battaglia 2020), and special issues entitled "Volumetric Sovereignty" (Billé 2019) and "Speaking Volumes" (Billé 2017). "Speaking Volumes" is paying attention to governance systems regulating dynamic processes of appropriating vertical space on land, in the air, and in the waters. Situated in this broad field of studying vertical appropriations, this chapter focuses on governing the underground as a zone of (potential) extraction. However, within underground volumes we also encounter water (aquifers), first within national boundaries and then beyond. After discussing subsoil governance, we move into deep waters, but only in relation to mineral extraction.

In most countries, with the notable exception of the USA, the state's authority to grant mining rights is based on a system in which the state is the ultimate owner of subsoil mineral wealth: the distinction between public and private coincides with the difference between subsurface and surface ownership. In literature on Latin America and Africa this national sovereignty over mineral wealth – so-called subterranean sovereignty – has either been framed as resource nationalism which opposes foreign capital (Childs 2016) or as an element of neoliberal strategies which facilitate alliances between the state and international companies at the expense of local groups' surface rights (Emel, Huber, and Makene 2011). Yet Jodi Emel and her co-authors argue that even where these alliances appear to be strong, empirical studies are required to analyse the way "subterranean sovereignty" and "layered rights" affect power relations between state authorities, companies, and local communities. Although

> sovereignty is negotiated at the level of the global-national scale, it is territorialized at the local scale. Thus, we must also keep attuned to the ways in which local populations living in the spaces of extraction are constantly interrupting state-capital sovereignty projects.
>
> *(ibid.: 73)*

In order to study hybrid forms of governance of underground mining, Luning and Pijpers (2017) propose the notion of in-depth geopolitics. Focusing on gold mining terrains in Ghana, it is clear that customary authorities play a major role as authorizers; the state may allocate formal permits to large-scale mining companies, but chiefs with control over surface land may facilitate access for small-scale miners even in areas within the formal permits. The article shows how in localized circumstances the layering of national and local authorities leads to hybrid governance arrangements (see also Cuvelier, Geenen, and Verbrugge this volume) which allow for diverse and sometimes parallel lines for negotiating access to the underground. Its approach towards in-depth geopolitics also considers how geological circumstances create room to manoeuvre for miners working at different scales. An industrial mining company may opt to allow small-scale gold miners to work at relatively shallow depths, while the company appropriates the mineral-rich volume located under the water table.

Another case of hybrid governance, situated in Queensland, Australia, presents interesting state–company–civil society dynamics, notably how these play out in public debates concerned with (the safety of) underground extraction. The study by Kim de Rijke et al. (2016) focuses on hydraulic fracturing for unconventional gas, most often called fracking, which is taking place in the underground aquifer system of the Great Artesian Basin. Even though the Australian government is authorizer of access to the underground in that it has the monopoly on handing out permits to mining companies, local farmers can provide access through their individual titles to surface land. Since these farmers extract underground water by drilling wells, they tend to be less opposed to the idea of tapping into the nether-world of resources than urban citizens who may be outright opposed to fracking. How can the public assess risks and be informed about the invisible situations under their feet which are key to their lives and survival? De Rijke et al. (2016) analyses how public and private experts play a role in the debates and decisions. The eth-nographic study of fracking in aquifers shows the complexities of state–society relations in Australia, the brokers of knowledge about invisible underground sys-tems, and the diversity among Australian citizens in their assessment of what fracking can or cannot do to the aquifer system.

In addition to studying the hybrid governance of the underground within nation states, anthropologists analyse discourses which attempt to frame national unity and uniformity through the medium of the underground. This builds on work on colonial Canada by Bruce Braun (2000) who points out that geological ways of seeing informed both ideologies of national wealth and policies promoting "rational" resource extraction. Proponents of this view – many of them geolo-gists – claimed that "proper" ways of extracting would lead to prosperity for the nation as a whole. David Kneas (2018) studies the case of the nation state Ecuador and how its image changed from a country with dire resource prospects to a nation with a future outlook of wealth based on copper: a true copper nation. Discourses portraying the great mineral potential of the country also tap into how the Andes as a whole has been constructed as a vast underground terrain of mineral wealth. Theoretically it is most interesting that the discourse on resource abundance is constructed on the basis of historical and territorial nationhood but configured as an interplay with subterranean features and structures which transcend the nation and its governance structures. The potentiality is assembled as a combination of the known and the unknown, the placed territoriality and the non-place and beyond. In recent years, these non-places and beyonds have become entangled in compe-tition over resources in deep sea exploration.

Interesting anthropological work also focuses on new scrambles for resources which are located beyond the current boundaries of nation states, notably under water. Since many forms of underground extraction are driven by scenarios of depletion, new "frontiers" are being explored. Deep sea mining (DSM) exploration is prominent and companies claim that "the ocean contains the largest known mineral inventories on earth" (Childs 2020: 196). This form of exploration makes it clear that a three-dimensional spatial approach needs to be supplemented by a fourth dimension, temporality. Both in legal terms and in their material form, undersea mineral wealth is marked by processes of becoming. Indeed, as Franck Billé (2017) points out, the making of resources depends on national and interna-tional authorities (UN) developing in-depth rules to govern the seas, a process tied to long-term competition between European sea powers since the seventeenth century. Until recently, this broadening of national ownership was confined to the water column, whereas the area below, including the seabed, ocean floor, and the subsoil thereof, were considered the common heritage of humanity. Currently, it is argued that this formulation is inapt now that technologies are beginning to facilitate the envisaging of extraction from the seabed. Thus, DSM presents the temporal process of making undersea resources legible and legal. Moreover, the material properties of deep-sea minerals accentuate temporal dimensions. As Childs (2020: 192) observes,

> the subterranean is not solid but porous, where the deep geological time of mineral formation is captured only at its very moment of becoming circumscribed as a resource and where matter changes state between molten rock, liquid water, mineral-rich gas and precipitated ore.

In order to break open UN regulations about the distinction between water columns and seabed, the deep-sea world is becoming the object of political technologies which script future needs in terms of "energy security". In securing the volume – through an underwater area, to use Elden's phrase (2013) – the four dimensions of space and time are articulated. Projects to access deep sea minerals require the interplay of "politics of time – anticipation, monitoring, adaptability, sustainability – and the politics of space" (Le Meur et al. 2018).

Paying attention to the relations between the subterranean and governmentality, in particular in spaces which are not – yet – territorialized, identifies interesting connections between technological developments and legal framing, and between governmental authority and other practices of claim-making. The intricate connections between politics and technologies of accessing the underground also indicate the importance of paying attention to work practices, both of and within mining companies.

5 Anthropology of mining companies and underground mining practices

Speculation is part of the politics of time inherent to the exploration stages of industrial mining projects: how much mineral wealth can be projected in future "resource making"? Anna Tsing's work on the Bre-X scam initiated anthropological interest in the role of speculation in assessing underground wealth (Tsing 2004). In order to attract investors, this Canadian junior mining company claimed to be sitting on mountains of mineral wealth, specifically alluvial gold, in Borneo, Indonesia; ensuing speculations turned out to be based on a massive hoax. Ferry (2020) describes how etymologically the term speculation has a history of being associated with the faculty of sight as well as the temporal quality of anticipation. In the first sense she identifies a shift from connotations of watchfulness and careful consideration towards a negative interpretation suggesting flimsy reasoning or conjecture. In the second sense, speculation facilitated the economic goal of investing with the aspiration of making profit. Both senses are important to our discussion of the underground as a potential domain for "making resources" by mining companies in the course of capitalist drives for profit and growth.

Oil and gas – and their geopolitical entanglements (Bridge and Le Billon 2017; Mitchell 2011) – are emblematic of these dynamics. Capitalism's craving for energy sources, in combination with scenarios of finitude – peak oil (Hughes 2017), end of oil (Roberts 2004) – make oil the subject of frantic histories and future speculation. Casting light on this situation, Mamdana Limbert (2015) explores projections, prognoses, and temporal horizons shaped by depletion calculation and reserves reporting. Studying experts involved in Shell's determination of proven reserves in their oil fields in Oman after a scandal broke in 2002 as a consequence of Shell's overstating these reserves by 40%, she shows that these processes of forecasting are shrouded in secrecy and received with suspicion by the inhabitants of Oman. Despite dire projections of depletion they have come to believe that what is being hidden is in fact more oil, not less.

Limbert's work illustrates how information which mining corporations make available in the public domain may be received with suspicion. No doubt, the specific interests of the mining industry play a role, but there is another element feeding into distrust. Information on the prognosis of reserves is especially difficult to verify due to their underground location; they are invisible unless mediated by visual tools and abstract calculative models. This invisibility makes citizens strongly dependent on professional forecasters as experts in seeing, determining, and calculating underground wealth.

As part of his work on copper mining in Ecuador (mentioned above), David Kneas (2016) takes a performative stance on the ways in which junior companies portray knowledge and assessments to investors in contrast to adjacent communities. For the first audience, they focus on subsoil wealth, for the second, they frame surface landscapes. Kneas shows how companies are imbricated in a complex relational field in which they provide various theatrical performances: for investors a spectacle of abundance, for local communities a landscape scarred by a history of overexploitation and deforestation. Moreover, he shows how a junior's enactment is entangled with the subsoil. Junior companies are precarious entities that live in a world of potential which often fails to materialize, an argument also developed in other work (Luning 2014; Luning and Pijpers 2017).

To this current anthropological work on mining companies, with its emphasis on the perspectives of experts and the higher echelons of staff within mining companies, are now added ethnographies focusing on working practices in the pits, on the actual extractors working on the ground and underground. It is important to note that, historically, the relation between higher echelons and miners has undergone dramatic changes due to technological innovations. Notably, new technologies of how to see and assess underground wealth have impacted on the division of labour between the engineers and managers, on the one hand, and the actual miners, on the other. An important aspect of the history of the industrialization of underground mining is the place of visual tools in scaling-up mining operations. In his study entitled *Seeing Underground*, Nystrom (2014) notes that practices of mapping the underground – the shape and location of the orebody and infrastructure to extract from it – only became part of mining operations in the second half of the nineteenth century. Before the engineers took over, the mine captains decided the strategy for tackling the ore, basing their judgement and planning on the ongoing process of extractive work. Thus, insights were produced as part of practical involvement in the process of underground extraction. While describing in great detail the tools and surveys engineers used in order to start drawing maps in drafting rooms in the aboveground office, Nystrom points out an interesting characteristic of the new division of labour between miners and planners of the work. In his search for maps in company archives, he noticed that they were all clean; they did not appear to have been used underground. The engineers had started to map and visualize above ground what was minable underground; consequently, work underground came to be seen in a mediated way. In the course of time most of the sophisticated visual tools for portraying underground wealth have become part and parcel of industrial mining. 3D block-modelling on computers, for example, allows mine managers to plan and strategize ways of extracting from the orebody.

In many cases mining companies – mainly companies extracting gold or coal – have shifted from underground mining to working bulk ore in open-cast mines. Rolston's ethnography on coal miners in the USA starts from the observation that "miners inhabit a work environment that they themselves create" (Rolston 2013: 582). She is interested in understanding how the maps made by engineers for guiding the carving out of pits translate into practice, equating the relationship between map and practice to one of design and material agency: every orebody has specific affordances which define possibilities for extracting from it and these material conditions act upon miners in their engagements with rock while working. This focus on how materiality works in practice leads to important insights into the relation between matter and substance. In the carving out of pits, orebodies have agency: rock happens. This has consequences for our understanding of the relation between maps, mining labour, and materiality. Maps are "representations of institutional decisions and plans, past and future" (Rolston 2013: 588), whereas miners are positioned as mediators between plans and their material manifestation, between visualizations of orebodies and embodied engagements in mining spaces.

6 Conclusion – from the underground rock to widening perspectives

This chapter started by briefly outlining the vertical turn which is taking place in different disciplines. It then turned to the political dimensions of studying the relentless race to the bottom and the top which characterizes capitalism. I suggest that the upsides of world making and the downsides of world breaking indicate the need to attend to exchange in the study of the vertical terrains of the social and subterranean. Exchange is a red thread which runs through the chapter to help rethink diversities in cosmologies and ontologies associated with labouring the underground. Tying the anthropology of the underground to the classical work of Polanyi creates a frame for coupling attention to labour and (underground) land. This is necessary for two reasons.

First of all, so the work of Yusoff teaches, it radicalizes questions about suffering, sacrifice, and the distinction between life and non-life; therefore, the decolonization of geology and perspectives on the underground should be pursued along these lines. By redefining extraction and the underground in more metaphorical ways (see e.g. Feige 2007), attention can be given to processes of social exclusion and dehumanization of people in histories of underground extraction. Jonkman's study on gold mining in Colombia (Jonkman 2021) uses the term "underground" metaphorically to draw attention to subaltern politics in the margins of the state. This political research agenda is furthered by the use of the term "undercommons" to stress that humanity is not in any way a self-evident, all-inclusive category (Harney and Moten 2013). This is an important message for the study of underground extraction and earthly matters more broadly, one which warns us that attention to ontologies cannot be equated with extending the notion of life beyond the human to multispecies. The way human energy has been extracted for the sake of extracting underground wealth turns on more complicated and painful classifications of how and whose life matters.

Secondly, returning to Polanyi's ideas helps further anthropological research into the cosmological aspects of resource making under capitalism. This ambition informed the second red thread in this chapter: visibility/invisibility in underground matters. This is central in my framing of different ways of knowing and modalities of exchange relations (based on whether or not the underground is associated with agency). Two lines of anthropological research deserve to be pursued here. The first line relates to the choice to stretch this chapter on the anthropology of the underground to extra-territorial "underwaters". This field needs to be connected more explicitly to anthropological studies which bring out the (cosmological) values of water. The sea is a domain for "extracting resources" of many kinds, including fish and pearls, and a rich anthropology on underwater worlds shows that cosmologies and exchange relations are key here too. One can think of Patricia Spyer's beautiful work on pearl divers in Eastern Indonesia (Spyer 1997), while African and African diaspora ideas related to the sea spirit "Mami Wata" (Van Stipriaan 2003) frame water as a cosmological domain which can give (or withhold) life and wealth. The second line was hinted at when recapitulating D'Angelo's attention to the importance of touch when visibility under water is hampered (D'Angelo 2019). We need to place sight within a more complete scope of the senses, an approach exemplified by ethnographies by Rosalind Morris and Felice Tiragallo, which explore relations between sight and the sonic (Morris 2008b; Tiragallo 2018). Morris' analysis of "the miner's ear" – blocked by the roaring noise of drilling but needed for hearing "moving rock" and other warning signals – evokes the tactile, sensual work experiences of miners, constantly at risk of underground accidents.

As a final comment, I want to return to the start of the chapter where I portrayed how colleagues in urban studies and geography situate extraction from the underground in wider relational and spatial contexts. I mentioned the link between mining and urbanization with the reference to cities as "inverted mines". Here I want to highlight Martin Arboleda's approach to "the planetary mine" (Arboleda 2020) in which he urges researchers to move away from singular attention to (underground) mining spaces as objects of study. He points out that current industrial mining practices depend on logistic infrastructures which operate on a planetary scale, describing the geography of movements and technologies from the sites of extraction to sites of manufacture, a trajectory mediated by autonomous trucks and shovels, a semi-automated train system, electro-refinery and smelting facilities, shipping containers, mega ports, and finally factories (Arboleda 2020). In his view this trajectory should lead to the "explosion" of the mine; scholars should look at territories of extraction on a planetary scale. This is an important research agenda to which anthropologists can contribute substantially with their patient and meticulous attention to how infrastructures are made and used in the daily, the routine, and the repetition of social practices.

This chapter demonstrates how the anthropology of the underground is part of a larger interdisciplinary conversation, albeit with distinct characteristics. It shows the importance of ethnographic work to general societal debates by bringing these debates to where they matter and materialize: tangible social practices of governance, organizational formats of mining companies, and work practices of experts and miners above and under the ground. It adheres to the anthropological stance of studying how ideas – which are always informed by cosmological perspectives and ontological postulates – materialize in mundane practices and daily routines. There is a strong anthropology which emphasizes that capitalism needs to be understood in its routines and repetitions, while current anthropological attention to infrastructure acknowledges that the daily work in pits and on oil rigs informs us of distributive agency (Rolston 2013) and forms of politics, such as corporate abdication of liability, racism, and disempowerment of workers (Appel 2015). By addressing all these facets of ongoing research trajectories, I have tried to show how the anthropology of the underground rocks.

Bibliography

Adey P. (2015) Air's affinities: Geopolitics, chemical affect and the force of the elemental. *Dialogues in Human Geography* 5 (1): 54–75.

Appel H. (2015) Offshore work. Infrastructure and hydrocarbon capitalism in Equatorial Guinea. In: Appel H., Mason A., and Watts M. (eds) *Subterranean estates: Life worlds of oil and gas*. Ithaca, NY: Cornell University Press, 257–273.

Arboleda M. (2020) *Planetary mine: Territories of extraction under late capitalism*, London: Verso.

Ballestero A. (2019) Underground as infrastructure? Figure/ground reversals and dissolution in Sardinal. In: Hetherington G. (ed.) *Infrastructure, environment and life in the anthropocene*. Durham: Duke University Press, 18–44.

Battaglia D. (2020) *Voluminous states: Sovereignty, materiality, and the territorial imagination*, Durham: Duke University Press.

Bélanger P. (2007) Underground landscape: The urbanism and infrastructure of Toronto's downtown pedestrian network. *Tunnelling and Underground Space Technology* 22 (3): 272–292.

Biersack A. (1999) The Mount Kare python and his gold: Totemism and ecology in the Papua New Guinea Highlands. *American Anthropologist* 101 (1): 68–87.

Billé F. (2017) Introduction: Speaking volumes. *Cultural Anthropology*. Available at: https://Culanth. Org/Fieldsights/1241-Introduction-Speaking-Volumes.

Billé F. (2019) Volumetric sovereignty part 1: Cartography vs. volumes. *Society and Space*. https://www.societyandspace.org/articles/volumetric-sovereignty-part-1-cartography-vs-volumes.

Bonnet D. (1988) *Corps biologique, corps social. Procréation et maladies de l'enfant en Pays Mossi, Burkina Faso*, Paris: Editions de l ORSTOM.

Braun B. (2000) Producing vertical territory: Geology and governmentality in late Victorian Canada. *Ecumene* 7 (1): 7–46.

Brechin G. (2006) *Imperial San Francisco: Urban power, earthly ruin*, Berkeley: University of California Press.

Bridge G. and Le Billon P. (2017) *Oil*, Hoboken: John Wiley & Sons.

Childs J. (2016) Geography and resource nationalism: A critical review and reframing. *Extractive Industries and Society* 3 (2): 539–546.

Childs J. (2020) Extraction in four dimensions: Time, space and the emerging geo(-)politics of deep-sea mining. *Geopolitics* 25 (1):189–213.

Chilvers J. and Burgess J. (2008) Power relations: The politics of risk and procedure in nuclear waste governance. *Environment and Planning A* 40 (8): 1881–1900.

Couling N. and Hein C. (2020) *The urbanisation of the sea. From concepts and analysis to design*, Rotterdam: NAi Publishers.

Crutzen P.J. (2006) The "anthropocene". In: Krafft T. (ed.) *Earth system science in the anthropocene*. Berlin: Springer, 13–18.

Crutzen P.J. (2016) Geology of mankind. In: Crutzen P.J. (ed.) *A pioneer on atmospheric chemistry and climate change in the anthropocene*. Berlin: Springer, 211–215.

Crutzen P.J. and Stoermer E.F. (2000) The "Anthropocene". *Global Change Newsletter* 41: 17–18.

D'Angelo L. (2016) Anthropology as storytelling: Fetishism and terror in Michael Taussig's early works. Polis. *Journal of Political Science* 4 (14): 77–89.

D'Angelo L. (2019) God's gifts: Destiny, poverty, and temporality in the mines of Sierra Leone. *Africa Spectrum* 54 (1): 44–60.

De la Cadena M. (2015) *Earth beings: Ecologies of practice across Andean worlds*, Durham: Duke University Press.

de Rijke K., Munro P. and de Lourdes Melo Zurita M. (2016) The Great Artesian Basin: A contested resource environment of subterranean water and coal seam gas in Australia. *Society & Natural Resources* 29 (6): 696–710.

Elden S. (2013) Secure the volume: Vertical geopolitics and the depth of power. *Political Geography* 34: 35–51.

Emel J., Huber M.T. and Makene M.H. (2011) Extracting sovereignty: Capital, territory, and gold mining in Tanzania. *Political Geography* 30 (2): 70–79.

Feige E.L. (2007) *The underground economies: Tax evasion and information distortion*, Cambridge: Cambridge University Press.

Ferguson J. (1999) *Expectations of modernity: Myths and meanings of urban life on the Zambian Copperbelt*, Berkeley: University of California Press.

Ferry E. (2020) Speculative substance: "Physical gold" in finance. *Economy and Society* 49 (1): 92–115.

Gellert P.K. (2005) For a sociology of "socionature": Ontology and the commodity-based approach. *Nature, Raw Materials, and Political Economy* 10: 65–91.

Golub A. (2014) *Leviathans at the gold mine: Creating indigenous and corporate actors in Papua New Guinea*, Durham: Duke University Press.

Gómez-Barris M. (2017) *The extractive zone: Social ecologies and decolonial perspectives*, Durham: Duke University Press.

Graham S. (2016) *Vertical: The city from satellites to bunkers*, London: Verso Books.

Harney S. and Moten F. (2013) *The undercommons: Fugitive planning and black study*, Wivenhoe: Minor Compositions.

Harris A. (2015) Vertical urbanisms: Opening up geographies of the three-dimensional city. *Progress in Human Geography* 39 (5): 601–620.

Hughes D.M. (2017) *Energy without conscience: Oil, climate change, and complicity*, Durham: Duke University Press.

Jacka J.K. (2015) *Alchemy in the rain forest: Politics, ecology, and resilience in a New Guinea mining area*, Durham: Duke University Press.

Jonkman J. (2021) *Underground politics. Socio-territorial relations, citizenship, and state formation in gold-mining regions in Chocó, Colombia*, Amsterdam: PhD Thesis. VU Amsterdam.

Khatua S. and Stanley W. (2006) Ecological debt: A case study from Orissa, India. In: Peralta A. (ed.) *Ecological debt: The peoples of the south are the creditors. Cases from Ecuador, Mozambique, Brazil and India*. Geneva: World Council of Churches, 125–165.

Kneas D. (2016) Subsoil abundance and surface absence: A junior mining company and its performance of prognosis in northwestern Ecuador. *Journal of the Royal Anthropological Institute* 22: 67–86.

Kneas D. (2018) From dearth to El Dorado: Andean nature, plate tectonics, and the ontologies of Ecuadorian resource wealth. *Engaging Science, Technology, and Society* 4: 131–154.

Le Meur P.Y., Arndt N., Christmann P. and Geronimi V. (2018) Deep-sea mining prospects in French Polynesia: Governance and the politics of time. *Marine Policy* 95: 380–387.

Limbert M. (2015) Reserves, secrecy, and the science of oil prognostication in Southern Arabia. In: Appel H., Mason A. and Watts M. (eds) *Subterranean estates: Life worlds of oil and gas*. Ithaca, NY: Cornell University Press, 340–352.

Luning S. (2012) Gold, cosmology, and change in Burkina Faso. In: Panella C. (ed.) *Lives in motion, indeed. Interdisciplinary perspectives on social change in honour of Danielle de Lame*. Brussels: Royal Museum for Central Africa, 9–26.

Luning S. (2014) The future of artisanal miners from a large-scale perspective: From valued pathfinders to disposable illegals? *Futures* 62: 67–74.

Luning S. and Pijpers R.J. (2017) Governing access to gold in Ghana: In-depth geopolitics on mining concessions. *Africa* 87 (4): 758–778.

Mayer P. (1961) *Townsmen or tribesmen: Conservatism and the process of urbanization in a South African City*, Cape Town: Oxford University Press.

Melo Zurita M.D.L., Munro P.G. and Houston D. (2018) Un-earthing the subterranean anthropocene. *Area* 50 (3): 298–305.

Mitchell J.C. (1959) Labour migration in Africa south of the Sahara: The causes of labour migration. *Bulletin of the Inter-African Labour Institute* 6 (1): 12–46.

Mitchell T. (2011) *Carbon democracy: Political power in the age of oil*, London: Verso Books.

Moore J.W. (2017). *Anthropocene or capitalocene? Nature, history, and the crisis of capitalism*. Oakland, CA: PM Press/Kairos.

Morris R. (2008a) *Notes on the underground: An essay on technology, society, and the imagination*, Cambridge, MA: MIT Press.

Morris R.C. (2008b) The miner's ear. *Transition* 98: 96–115.

Nash J. (1979) *We eat the mines and the mines eat us: Dependency and exploitation in Bolivian tin mines*, New York: Columbia University Press.

Nystrom E.C. (2014) *Seeing underground: Maps, models, and mining engineering in America*, Reno: University of Nevada Press.

Pijpers R.J. and Eriksen T.H. (2018) *Mining encounters: Extractive industries in an overheated world*, London: Pluto Press.

Polanyi K. (1944) *The great transformation: The political and economic origins of our time*, Boston: Beacon Press.

Povinelli E.A. (2016) *Geontologies: A requiem to late liberalism*, Durham: Duke University Press.

Povinelli E.A. (2019) The urban intensions of geontopower. *e-flux*. Available at: https://www.e-flux.com/architecture/liquid-utility/259667/the-urban-intensions-of-geontopower/.

Richardson T. and Weszkalnys G. (2014) Introduction: Resource materialities. *Anthropological Quarterly* 87 (1): 5–30.

Roberts P. (2004) *The end of oil. The decline of the petroleum economy and the rise of a new energy order*, London: Bloomsbury.

Rolston J.S. (2013) The politics of pits and the materiality of mine labor: Making natural resources in the American West. *American Anthropologist* 115 (4): 582–594.

Rosen L.C. (2020) *Fires of gold: Law, spirit, and sacrificial labor in Ghana*, Berkeley: University of California Press.

Ross M.L. (1999) The political economy of the resource curse. *World Politics* 51 (2): 297–322.

Sheller M. (2014) *Aluminum dreams: The making of light modernity*, Cambridge, MA: MIT Press.

Sosa M. and Zwarteveen M. (2012) Exploring the politics of water grabbing: The case of large mining operations in the Peruvian Andes. *Water Alternatives* 5 (2): 360–375.

Spyer P. (1997) The eroticism of debt: Pearl divers, traders, and sea wives in the Aru Islands, eastern Indonesia. *American Ethnologist* 24 (3): 515–538.

Taussig M. (1980) *The devil and commodity fetishism in South America*, Chapel Hill: University North Carolina Press.

Tiragallo F. (2018) Tunnels of voices. Mining soundscapes and memories in South West Sardinia. *Ethnologia Polona* 39: 11–29.

Tsing A. (2004) Inside the economy of appearances. In: Amin A. and Thrift N. (eds) *The Blackwell cultural economy reader*, Oxford: Blackwell Publishing, 83–100.

van de Camp E. (2016) Artisanal gold mining in Kejetia (Tongo, Northern Ghana): A three-dimensional perspective. *Third World Thematics: A TWQ Journal* 1 (2): 267–283.

Van Stipriaan A. (2003) Watramama/Mami Wata: Three centuries of creolization of a water spirit in West Africa, Suriname and Europe. *Matatu* 27 (1): 321–337.

Weizman E. (2002) The politics of verticality. *Open Democracy*. Available at: https://www.opendemocracy.net/en/article_801jsp/.

Werthmann K. (2003) Cowries, gold and "bitter money": Gold-mining and notions of ill-gotten wealth in Burkina Faso. *Paideuma* 67: 105–124.

Yusoff K. (2015) Queer coal: Genealogies in/of the blood. *PhiloSOPHIA* 5 (2): 203–229.

Yusoff K. (2018) *A billion black Anthropocenes or none*, Minneapolis: University of Minnesota Press.

12

WATER AND CONFLICT

Fabiana Li and Teresa A. Velásquez

1 Introduction

Conflicts over resources have attracted considerable attention in recent years, giving rise to both scholarly and popular debate about the ways that extractive projects affect communities and other actors engaged with their operations. The frequency and intensity of these conflicts in some parts of the world have put controversies over extraction at the forefront of regional and national politics. These conflicts also have international reverberations, as they often involve expansive global networks that include transnational corporations, international organizations, and solidarity activists. Most recently, concerns over the environment – and water, specifically – have changed the nature of activism and corporate responses to these conflicts, and have also brought about new forms of academic analysis. Indeed, while conflicts over resource extraction revolve around a confluence of various issues and respond to a diversity of grievances, some of the most emblematic conflicts in recent years have highlighted the risks of pollution and the destruction of water resources that can accompany mineral extraction (Bebbington and Williams 2008; Kirsch 2014; Li 2015). In this chapter, we explore the significance of the environment (which cannot be separated from social, political, and economic concerns) in popular and academic debates around mining, and the role of water as a substance that becomes drawn into activism, negotiation, social relations, scientific studies and expert reports, political discourse, and representations of conflict.

The prominence of water in these conflicts is in part a consequence of geography: mining projects are often located at the headwaters of watersheds (in the case of open-pit gold mining) or near rivers and lakes (as with underground and artisanal mining) and interact with groundwater systems. The centrality of water in many recent conflicts also results from a combination of social and political factors:

DOI: 10.4324/9781003018018-12

the global circulation of environmentalist discourses, increased awareness about the effects of water pollution and water scarcity, and greater visibility of environmental and indigenous movements in national politics and international media.

We begin this chapter with a section that contextualizes water-related conflicts in broader mining struggles and explore the political and economic shifts in countries where mineral disputes have proliferated in recent decades. Focusing on the rise of the new geographies of extraction, we trace how the expansion of the mining frontier produces tensions among different actors as they compete for the power, authority, and right to govern resources. Acknowledging that different scales of mining may engender different types of conflict, we comparatively examine artisanal and small-scale mineral extraction (ASM) and large-scale mining, drawing attention to their relevance for the territorial dynamics that define the new geographies of extraction. We then explore "water infrastructure", a term we use to describe the material, social, and political relationships that shape water conflicts in mining regions. We suggest that water is an important point of contention that influences forms of engagement and resistance. We then examine how anti-mining activists increasingly turn to science and citizen referendums to defend a variety of claims, suggesting that the defence of water and landscapes in mining conflicts are inextricably connected to the defence of rights and sovereignty. In the concluding section, we focus on future areas of research for water and mining-related conflicts.

2 New geographies of extraction

A broad survey of the global mining industry from the early 2000s onwards indicates a process of diversification and increased investment that expanded extraction into new areas and intensified extraction in traditional oil and mineral-producing countries (Bebbington 2009). In many parts of the world, this shift was accompanied by neoliberal economic and legislative reforms that encouraged foreign investment, often as part of structural adjustment programmes promoted by the International Monetary Fund and the World Bank (Ferguson 2006; Fraser and Larmer 2010; Sawyer and Gomez 2012). The opening up of the economy made the extraction of raw materials, including minerals and metals, central to national development plans. However, the expansion of extractive industries seems to transcend the political ideologies (usually defined as "left-" or "right-leaning"), as governments of various political stripes have supported "neoextractive" models that promote resource-led development as fundamental to a country's economic progress (Canterbury 2018; Farthing and Fabricant 2019; Gudynas 2010; Roder 2019).

Across the globe, people have historically engaged in informal mining activities, but global political and economic shifts have given rise to an explosion in the number of people reportedly engaged in artisanal and small-scale mining (ASM) in the last two decades (Lahiri-Dutt 2018). ASM refers to a range of labour arrangements, production methods, and technologies but is usually characterized by low-tech and labour-intensive methods of extraction (Lahiri-Dutt 2018; Verbrugge and Geenen 2020). ASM activities may occur on the margins of the law and often exist dynamically with a range of other forms of livelihood, such as agriculture (Pijpers 2014). A boom in commodity prices in the 2000s, alongside legal reforms to attract global mineral investment, have pushed people around the world into ASM activities (Marston 2017; Yankson and Gough 2019). At the same time, record-high metal prices incentivized companies to develop projects previously deemed to be economically unfeasible due to low-grade ore deposits, costly methods of extraction, and difficult locations. With favourable legal-economic frameworks in place to facilitate resource extraction, companies could invest in large-scale projects using technologies such as open-pit cyanide leach mining (Verbrugge and Geenen 2020). Where this activity overlapped with ASM, new conflicts between artisanal miners and global mining companies emerged over operating permits and land claims, sometimes culminating in the forced dispossession of artisanal miners (Hilson and Potter 2005).

These dynamics are well illustrated in Ghana's gold mining sector, where over one million miners operate without licences (Tschakert 2009). In Ghana, the government and media propagated a public discourse of artisanal gold miners as reckless to the environment, "violent criminals", a "menace", and a "threat" (Hilson and Yakovleva 2007: 104; Tschakert 2009: 712). These representations emerged from the use of mercury in artisanal and small-scale gold mining (ASGM) operations and competing claims over mineral-rich landscapes (Tschakert and Singha 2007), culminating in a 2017 country-wide ban on all ASGM activities. This had particularly devastating effects on the livelihoods of women who depend upon ASGM activities (Hilson 2017; Kumah et al. 2020; Ntewusu 2018). Negative representations associated with artisanal gold mining were used in public arguments for the development of large-scale mining by mobilizing false dichotomies of legal/ illegal mining; environmentally responsible/reckless mining; and formal/informal mining – a problematic construction considering that these dichotomies are co-produced in mining conflicts (Hilson and Yakovleva 2007; see also Hausermann et al. 2018; Moore and Velásquez 2013). Artisanal miners are often pivotal for the development of large-scale mining as "junior" mining companies "free ride on artisanal miners as an index for promising patches," and then dispossess them once they stake their claim (Luning 2014: 70). However, governing at the site of extraction is a complex arrangement that can depend upon the geology of the mineral deposits and the history of occupation by artisanal miners that sometimes makes for a relationship of co-habitation rather than dispossession (Luning and Pijpers 2017).

The expansion of large-scale operations in the resource frontier opens up ques-tions of power and authority, illustrating that conflicts over who has the right to work, to access resources, and to live in these landscapes are complicated by over-lapping claims, the influx of immigrants, and the relationships between ASM and global mining companies (de Theije and Bal 2010; Tubb 2020). Groups may stra-tegically appropriate discourses of tradition and history to legitimize authority over certain areas, reconfiguring power dynamics and social networks (Grätz 2009). With sustainable development and industrial agriculture as new economic priorities, various concession holders (miners, indigenous communities, agri-busi-nesses, eco-tourism operators, alongside armed groups) compete for authority to govern the extractive frontier, especially when the central government has little or no capacity to regulate these "distant" places (de Theije and Salman 2018; Urán 2018). Cuvelier (2013) notes that in the Democratic Republic of Congo, despite government efforts to dismantle the shadow networks that support so-called "conflict minerals", local powerbrokers with past links to armed groups use vio-lence, corruption, and the manipulation of official and unofficial rules to govern extraction, creating a hybrid political order in which state and non-state actors compete for authority in resource-rich areas.

Contributing to the territorial dynamics of large-scale mining expansion are international agreements, norms, and discourses that put a spotlight on indigenous rights and the environment. International human rights norms play an increasingly important role in asserting resource control in mining-related conflicts, but they come with their own political risks. Anti-mining communities have turned to the United Nations Declaration on the Rights of Indigenous Peoples (UNDRIP) as a mechanism to uphold their rights to free, prior, and informed consent (FPIC). However, as cases for indigenous consultation are brought before international human rights courts, critical questions emerge: who qualifies as an indigenous group, and how is indigeneity defined in the absence of certain cultural markers (Kirsch 2018)? In Sweden, mining companies invoke narrowly defined "biological and cultural regimes" to argue that Sami reindeer herders are not "different enough" from the rest of the population to merit consultation rights (Lawrence and Moritz 2019: 46–47). When the FPIC process is controlled by governments amenable to extraction, they often select indigenous leaders that support mining or treat minority opinions as the majority (Camba 2016). The recognition of rights and territorial governance can also be complicated by the existence of plural regimes where different criteria are used to identify who has authority to make decisions over lands and what property ownership entails (Horowitz 2002).

The expansion of large-scale metal mining since the 2000s has also been accompanied by the formation of new social movements that are often character-ized by multiethnic and heterogeneous coalitions. Some movements may involve indigenous and peasant groups but have varied discursive frameworks, demands, and political proposals. Other movements find indigeneity to be a politically pro-ductive identity category that helps them articulate a complex set of demands that may include mining on their own terms (McDonell 2015). Different types of mining project have faced a variety of resistance and negotiation strategies that include protests, referendums, and international solidarity campaigns. In cases where water has become a central concern in activism, companies and the state have responded in ways that seek to channel and redirect their activism towards participatory efforts, such as community water monitoring programmes (Himley 2014). Increasingly, there is a trend towards the criminalization of mining-related activism that ranges from legal harassment to physical attacks (Billo 2017; Rasch 2017).

In the sections that follow, we explore how water shapes different geographies of extraction, particularly when water's material, social, and political dimensions are implicated in mining-related conflicts. We then trace emerging movements and networks that engage expert knowledge and legal strategies to defend water. In some recent cases, the presence of water resources (used for household activities, irrigation, livestock, or artisanal mining processing) has influenced the dynamics of the conflicts that have emerged, including what actors they involve and how they mobilize social movements and resistance strategies. These movements may reference social movements of the past, but as recent studies have shown, they also diverge from them in important ways. Focusing on water in mining conflicts provides a lens through which to examine issues of contestation, negotiation, and resistance.

3 Water conflicts

In this section, we describe how water organizes and enfolds people, the environment, and global corporations into conflict. To do so, we use the term "water infrastructure", a concept that refers to the way that water has become a terrain of conflict through which negotiation, engagement, and resistance to mining occurs. Water infrastructures encompass the material, social, and political relations that organize mining-related conflicts. How and where water flows, the quality and quantity of water, and the valuations of water give rise to different types of conflict surrounding metal mining. Depending on the type of mining, the location of the operations, and the bodies of water affected, resistance to extractivism has also taken different forms. To understand the centrality of water in mining conflicts, we focus on five key themes that have emerged in the academic literature and through the actions of grassroots actors mobilizing in response to mining activity: 1) water is a biophysical material whose quality and quantity are contested, 2) water is the focus of negotiation and compensation, 3) water is a relation, 4) water is multiple, and 5) water is a catalyst for political action.

Water is a biophysical material whose quality and quantity are contested

In recent mining conflicts, different geographies of water, technologies of production, and kinds of mining operation have played a role in shaping pollution concerns. Large-scale mines tend to be located at the headwaters of a watershed, affecting communities downstream. Open-pit mining requires draining water from mining pits in order to mine safely and uses large quantities of water in its operations. The diversion of water streams and irrigation canals to make way for a mine's operation, or to avoid contamination from mine runoffs, can also affect water availability in neighbouring communities. In Peru, the pollution of rivers and streams, the reduction of freshwater for household consumption and irrigation, and other changes in the water cycle have generated resistance to new mining projects.

Water quality and quantity have emerged as a focal point of recent controversies, as in mobilizations that sought to protect aquifers and mountain lakes from large-scale mining activity (Bos and Grieco 2018). Water was not only central to the demands of protestors, but it helped to mobilize people through their connections to water and irrigation boards, shared sources of water, or exposure to a mine's effects on pastures and soil (Li 2016; Rasmussen 2015). In spite of the ways that mining activity interacts with water resources, some mining corporations maintain that they operate with the highest social and environmental standards in the industry, which they aim to show through their Corporate Social Responsibility initiatives (Dolan and Rajak 2016; see also Shever this volume).

In contrast, in some ASM operations, the accumulation of tailings in wetlands, lakes, rivers, and other bodies of water caused erosion and increased sediment loads, jeopardizing the livelihoods of farmers and migrants living on or near riverbanks, forcing them to seek other sources of water for bathing, irrigation, and drinking (Beavis and McWilliam 2018). Moreover, in gold mining, mercury evaporates in the amalgamation process and can settle back into local streams, further affecting communities (Beavis and McWilliam 2018).

Water is the focus of negotiation and compensation

Water can be central in negotiations between companies, communities, and the state. In Papua New Guinea, for example, the release of effluents into the Ok Tedi and Fly rivers had a direct impact on people's livelihoods (Kirsch 2014), leading to claims for compensation, calls for closure, and legal challenges. Water can influence relations of power in negotiations, compensation agreements, and lawsuits, and being part of an affected watershed or an association of water users can have certain advantages when it comes to securing compensation or other benefits. In Peru, water users have received payouts and work contracts as compensation for damages to their irrigation canals and for reduced water quantities for irrigation resulting from the Yanacocha mine operations (Li 2015). In some of these cases, community leaders negotiated directly with the company, leading to tensions and disagreements among community members. For example, community members may have different interests based on their degree of dependence on farming, their involvement in maintaining water infrastructure, and their notions of rights that govern community water use (Sosa and Zwarteveen 2012).

Communities often depend on mining companies for employment, yet modern mines typically do not create enough jobs to meet the local demand. To mitigate this problem, some companies have encouraged the creation of local businesses, or micro-enterprises, subcontracted to provide services that are often tied to water – for example, the construction and maintenance of canals or the installation of irrigation systems (Li 2015). Corporate investment in communities also involves the provision of water-related services, such as potable water for rural households. While people depend on water resources that are threatened by resource extraction, mining companies paradoxically take on the role of regulating and managing those same resources – even as their operations put them at risk (Li 2015).

People's dependence on water resources for growing crops and raising livestock does not preclude their desire for employment and economic opportunities to supplement their livelihoods, and can lead community members to cooperate with mining companies or oppose their operations. For this reason, people's relationships with extractive industries often oscillate between resistance, partnership, and collaboration. In the Pacific Northwest of North America, an adversarial relationship between a mining company and an activist environmental coalition turned into one of "critical collaboration" (Smith Rolston 2015), where the grassroots coalition worked with the corporation to reach an environmental agreement. Critical collaboration can bring both opportunities and pitfalls, as the concessions and compromises that have to be made may not meet the expectations of the various actors involved. The tensions, changing positions, and contradictions that characterize community responses to mining activity should not be overlooked in the analysis of resource conflicts.

Water is a relation

Water connects people, places, and things, reconfiguring our understanding of human–nonhuman relations. The building of water installations for extraction also produces and enlivens new conflicts. In Mongolia, the Oyu Tolgoi copper mine built a network of pipelines, boreholes, and water pumps in the Gobi Desert that not only diverted water away from nomadic herder wells to the mine, but also decontextualized water from its cultural, environmental, and historical significance (Jackson 2018). The company built an underground pipe to divert the Bor Ovoo spring to a new location. Recreating the spring in a new site sparked opposition as locals rejected the "western ontological split between water and land that assumes economic value outweighs non-economic relationships to the waterscape" (Jackson 2018: 348). Facilitated by changes in both the mining and water laws, the case of Oyu Tolgoi resonates with other cases where mineral expansion entails the production of water as an inalienable commodity separate from culturally and historically informed notions of territory (Bebbington et al. 2010). In other parts of the world, water is part of a wider set of relationships. For Andean farmers, for example, bodies of water are connected to each other, and disruptions in one cause a break in the relations that maintain human and nonhuman life. Protecting sources of water is also about nourishing relations between people and the land that feeds them, and maintaining reciprocal relationships of care (Paredes Peñafiel 2020; Salas Carreño 2016). Recent conflicts have made these human–nonhuman relations feature more prominently in the political sphere and academic research.

Water is multiple

In the literature, water's multiplicity refers to the different meanings and values attached to it. In Chile's Atacama Desert, indigenous communities have challenged the commoditization of water (privatized in the 1980s), exposing competing systems of meaning and value (Babidge 2016; Bottaro et al. 2014; Carrasco 2016). Water is embedded in the reciprocal relations that sustain community irrigation practices and whose "value is rooted in the affective, communicative, and material assemblage between human and water flows" (Prieto 2016: 31). Across the globe, neoliberal mining reforms were accompanied by new water laws to formalize and standardize water rights, which privileged particular valuations in water access and governance "aimed at the commensuration of different waters to make their comparison and exchange possible" (Sosa et al. 2017: 217). The reforms aimed at producing "modern water" (Linton 2010) allowed for water to be physically and legally diverted to large-scale mining operations, which can result in the erosion of the cultural and epistemological complexity of customary notions of water management and rights (Akiwumi 2015; Sosa et al. 2017). For many rural communities, water access and management are governed through customary regimes that involve the exchange and sharing of water resources, which are not always formalized or recognized by national law. New water laws can create a market where water rights are exchanged and sold in ways that privilege economically powerful actors, such as mining companies, while dispossessing indigenous communities who might sell their rights with incomplete information (Budds 2010) or who may hold different valuations of water.

Water is a catalyst for political action

Water conflicts have mobilized diverse actors, created new alliances, and produced new forms of politics. These different ways of apprehending water have shaped recent conflicts, academic analysis, and political discourse around resource extraction. Glaciers, aquifers, mountain lakes, and other bodies of water, have become powerful symbols associated with the "defence of life" in protests against resource extraction. Sometimes, protestors allude to "sacred" or spiritual connections to these bodies of water (Valladares and Boelens 2019), referencing the ontological struggles that have come to frame mining conflicts (De La Cadena 2010; see also Acuña 2015). In Ecuador, people have defended their watersheds as the embodiment of the *pachamama* – a life-generating entity that sustains human and non-human lives – by holding outdoor rituals as a political strategy that draws rural and urban groups together in the "defence of life" and reconfigures the meanings and values attributed to the watery landscapes slated for extraction (Velásquez 2018). In India, where forests and waterways are entangled with indigenous subsistence, identity, and spirituality, indigenous women have also been particularly active in questioning the extractive development model that has often left poverty in its wake (Novo et al. 2018).

4 Emerging movements and networks to defend water

In this section, we continue our focus on the infrastructure of mining conflicts by deepening our analysis of the role that water plays in shaping political action. An expansion of the extractive frontier coincides with a reconfiguration of popular politics. The emergence of new coalitions and networks that span local, regional, and national scales build upon an earlier generation of social movements in so far as they seek to democratize development (Escobar and Alvarez 1992). However, they do not privilege identity-based rights as a primary demand for their activism even if they do intersect with indigenous and gender politics. Instead, their focus on expertise and the legal system shapes the formation of heterogeneous coalitions and alliances. We expand upon political ecology approaches that centre questions of social movements, the environment, and the democratization of development (Peet and Watts 2004) by illustrating how water-related concerns enable the formation of broad-based coalitions in the resistance to mineral extraction. In this section we explore the ways that movements coalesce to defend watersheds and landscapes by drawing upon expert knowledge and the law, highlighting the possibilities, risks, and challenges that emerge when movements dispute extraction through idioms of power.

Knowledge, evidence, and activism

Today's conflicts do not only involve direct "stakeholders", but enroll an international network of actors ranging from anthropologists (Coumans 2011a; Kirsch 2014; see also Bainton and Skrzypek this volume) to journalists, documentary filmmakers, environmental consultants, and hydrology specialists. The way that knowledge travels locally, around the world, and back again has the potential to solidify and strengthen the claims of social movements and their allies. The speed of information sharing in a globalized world also means that these knowledges are always fluid and easily destabilized, as new information shapes and reshapes mining debates and local responses to mining activity.

Amidst this proliferation of information, the potential risks of mining on human health and the environment have elevated the role of "experts", an encompassing term that anthropologists have used to include farmers, indigenous elders, scientists, international consultants, and others whose knowledge shapes the development of mining conflicts (Zhouri 2017). Studies of resource extraction have examined the role of science and expertise in controversies around mining (Graeter 2017; Macintyre and Foale 2004; Velásquez 2012). As a requirement for mining development, Environmental Impact Assessments (EIAs) and other techno-legal mechanisms have become a focal point of some conflicts and the efforts of NGOs, activists, and corporations to support their respective cases for or against the mine (Horowitz 2010). The burden of "proof" that communities have experienced contamination or other detrimental effects of mining rests on NGOs and other grassroots actors, who often lack the kind of funding and personnel to carry out environmental impact studies. In most cases, such efforts cannot match the funds and public relations activities that corporations devote to creating and disseminating their own environmental data.

In spite of the challenges, activists have used technical and scientific arguments and studies to support their opposition to mining projects (Eriksen 2018). This requires demonstrating the presence of pollution or other impacts that are not always tangible or readily evident. In some conflicts, water has provided a medium through which these effects become tangible (as with the change of colour, taste, or flow levels) and visible to an outside audience (through the publication of studies or news stories). Water provides a common language that can bridge local understandings and technical data (such as pH levels and Maximum Permissible Limits for toxic substances), and support the claims of communities (Himley 2014; Perreault 2013). However, water can also create (or exacerbate) divisions between groups with different interests, strategies, and objectives, such as environmentalists and local communities. Controversies involving water can also lead to contestation over the legitimacy of various forms of knowledge – variously labelled as traditional, indigenous, local, scientific, or technical knowledge (Long et al. 2013). When these competing knowledges are used to support, energize, and amplify the voices of activists and local people, they can pose a challenge for corporations trying to create a dominant narrative around the benefits of mining activity (Lassila 2018). At other times, the discrepancies between different forms of knowledge can sow doubt and confusion, leading to fragmented social movements and distrust among different constituencies.

Citizen referendums

In the last two decades, activists have turned to referendums as a strategy to defend watersheds from mineral extraction. The effects of mineral extraction on waterways, particularly the quality and quantity of drinking water sources and irrigation supplies, are central concerns that have culminated in citizen referendums. These referendums defend the "community as a legitimate scale for decision-making", redistributing political power to those most affected by mineral extraction (Urkidi 2011: 556). In Greenland, the lifting of a national ban on uranium mining prompted public outcry over the absence of citizen participation in development decision-making and sparked calls for a referendum (Nuttall 2013). The popularity of referendums is facilitated by networks that help "circulate information, experiences and strategies, and promote the mobility of activists to learn and share experiences among communities" (Walter and Urkidi 2017: 275).

Referendum movements are shaped by a politics of anticipation as they entail an effort to stop mineral extraction before environmental and health hazards become apparent or before mining companies enter with promises of jobs and social benefits (Kirsch 2014). Building a social movement around the "politics of time" provides an antidote to the "politics of resignation", in which corporate power appears as inevitable or unchangeable (Kirsch 2014). Mining disasters, such as in the mountains of Marinduque, Philippines, where Placer Dome dumped 200 million tons of metal-leaching waste in Calancan Bay between 1975 and 1991 and severely affected the livelihood of 12 fishing villages, serve as cautionary tales of the harm that may come with mineral projects (Coumans 2011b). As Coumans writes, "Fifteen years ago communities were more likely to be receptive, or at least resigned, to the prospect of hosting a large-scale mine ... Now, potentially affected community members, even in remote locations, are increasingly opposing mining before it starts" (Coumans 2011b: 115). In the Philippines, at least 10 provinces and 32 municipalities have passed moratoriums to ban large-scale mining (Coumans 2011b). While moratoriums differ from citizen referendums, they are often inter-linked as coalitions of citizens pressure government officials to adopt bans on large-scale mining (Bebbington et al. 2019).

The concerns, content, and form of the citizen referendums are often shaped by the kinds of environment where mining projects are located, and the relationship between people and those environments. In desert and arid areas, such as Candarave, Peru, referendum movements raised fears that mining would reduce water quantity or pollute already scarce water resources, thereby deepening the water crisis that forced migrants to leave the area (Walter and Urkidi 2017). A preoccupation with water shaped the referendum question where people voted to prohibit the use of underground and superficial waters for mining. In water-rich areas, such as the *páramos*, groups also fear a reduction of water quantity and quality, but the referendum question and processes are also shaped by their distinct relationship to the upland moorlands. Diverse citizen groups in the department of Santander, Colombia, led a referendum movement to declare *páramo* ecosystems as a collective good and in the public interest (Hincapié 2017). In Ecuador, an organization of community drinking water boards convoked a *consulta* following the democratic procedures that govern the institution based on the number of water rights held by a household (Riofrancos 2017). The question posed, "Are you in agreement with mining activity in the wetlands (*páramos*) and watershed of Kimsacocha?" (Riofrancos 2017: 684), referenced the ways that community water boards, through their many years of activism, had reframed the IAMGOLD (now INV Metals) "Quimsacocha" mine project into "Kimsacocha", a life-generating entity (Velásquez 2018).

Although citizen referendums often culminate in a single question (or two), they are motivated by processes that open up many other questions: "Without water, what will we do? How will our animals survive?" (Nelson 2013: 3). Referendums invite people to weigh not only the risks and equivalencies of mining, "How much is your river worth?" (Middeldorp et al. 2016: 934), but also "the political, economic and legal powers of mining companies" (Nelson 2013: 7). Water-related concerns often hold referendum movements together, but these processes are also shaped by broader histories of territorial conflicts and common experiences of social and economic marginalization (Hincapié 2017; McNeish 2017; Middeldorp et al. 2016). Referendums can link the demand for self-determination with broader critiques of colonialism, state racism, and development (Copeland 2019; Fultz 2016; Rasch 2012; Yagenova and Garcia 2009). However, indigenous sovereignty cannot be reduced to an anti-mining position. In New Caledonia in the southwest Pacific, a referendum on national independence held significance for Kanak self-determination where engagement with nickel extraction was perceived as a path to indigenous "voice and visibility" in the global system (Gentilucci 2019).

By raising questions over who has the authority to decide if, when, and where mining activities occur, referendum movements continue to expand upon the meaning and practices of public involvement in extractive projects (Ballard and Banks 2003); however, they are not always legally binding, making them an important symbolic strategy to protect watersheds and landscapes.

5 Conclusion

The expansion of large-scale mining projects, favourable economic conditions intended to attract foreign investment in mining, and the proliferation of artisanal and small-scale mining (ASM) alongside corporate mining interests have come to define the new geographies of mining conflicts. Moreover, and particularly relevant in light of this chapter, different waterbodies and types of water are implicated in the extraction of mineral deposits that were previously deemed marginal or not economically feasible, giving shape to various forms of engagement, conflict, and resistance. Subsequently, water gets invoked in mining conflicts in different ways depending on the location and technical specificities of the project: tailings dumped into rivers and oceans, runoff from waste-rock deposits, or the collapse of dams near mining operations produce different risks and concerns for local populations. Their vicinity to a project, and their use of resources that may be compromised by mining operations, can shape the types of response to extractive activity.

Whether a mining project diverts irrigation water from an aquifer or compromises a city's water supply, heterogeneous coalitions have taken up various methods and strategies to defend bodies of water, and often under the banner of protecting human and non-human life. The various meanings attributed to water, different ways of knowing and understanding pollution and loss, and competing claims to water rights are the underlying tensions that produce resource conflicts. As mining companies enter with promises of environmental care and sustainability, often at the expense of ASM operators, the realities of pollution in high-profile mining projects are a warning for many people of what mining may bring. In anticipation of mining-related disasters, social movements determined to keep their watersheds and landscapes "free of mining" are turning to local referendums and pressuring governments to ban large-scale mining.

As mineral investment continues to expand, deep sea mining is perhaps the "next frontier" of extraction. Deep sea deposits are comprised of polymetallic nodules that hold nickel, cobalt, copper, and magnesium. Similar to other bodies of water, oceans hold socio-cultural significance for coastal communities, providing food security and economic livelihoods that are entangled with identity, and symbolic and spiritual meaning (Chin and Hari 2020). Activists and scientists alike are concerned about the risks and lack of knowledge about the impacts of deep sea mining that range from the anticipated loss of biological diversity to the unknown chemical reactions and bioaccumulation released by mining waste and its effects on the deep sea (Chin and Hari 2020). Technologies like deep sea mining, along with the continued global demand for resources, have the potential to expand resource frontiers into uncharted waters. However, these developments will not go uncontested. Coalitions to protect water bodies from extraction are likely to continue to emerge in the coming years, adding new geographies of resource conflicts to the map.

Bibliography

Acuña R.M. (2015) The politics of extractive governance: Indigenous peoples and socio-environmental conflicts. *The Extractive Industries and Society* 2 (1): 85–92.

Akiwumi F.A. (2015) Analyzing Sierra Leone's water reform efforts: Law, environment, and sociocultural justice issues. *Politics, Groups, and Identities* 3 (4): 655–659.

Armah F.A., Luginaah I. and Justice O. (2013) Artisanal small-scale mining and mercury pollution in Ghana: A critical examination of a messy minerals and gold mining policy. *Journal of Environmental Studies and Sciences* 3 (4): 381–390.

Babidge S. (2016) Contested value and an ethics of resources: Water, mining and indigenous people in the Atacama Desert, Chile. *The Australian Journal of Anthropology* 27 (1): 84–103.

Ballard C. and Banks G. (2003). Resource wars: The anthropology of mining. *Annual Review of Anthropology* 32 (1): 287–313.

Beavis S. and McWilliam A. (2018) Muddy rivers and toxic flows: Risks and impacts of artisanal gold-mining in the riverine catchments of Bombana, Southeast Sulawesi (Indonesia). In: Lahiri-Dutt K. (ed.) *Between the plough and the pick: Informal, artisanal and small-scale mining in the contemporary world.* Canberra: ANU Press, 295–310.

Bebbington A. (2009) Latin America: Contesting extraction, producing geographies. *Singapore Journal of Tropical Geography* 30 (1): 7–12.

Bebbington A., Fash B. and Rogan J. (2019) Socio-environmental conflict, political settlements, and mining governance: a cross-border comparison, El Salvador and Honduras. *Latin American Perspectives* 46 (2): 84–106.

Bebbington A., Humphreys-Bebbington D. and Bury J. (2010) Federating and defending: Water, territory and extraction in the Andes. In: Boelens R., Getches D. and Guevara Gil A. (eds) *Out of the mainstream: Water rights, politics and identity.* London: Earthscan, 307–327.

Bebbington A. and Williams M. (2008) Water and mining conflicts in Peru. *Mountain Research and Development* 28 (34): 190–195.

Billo E. (2017) Mining, criminalization, and the right to protest: Everyday constructions of the post-neoliberal Ecuadorian state. In: Leonard L. and Grovogui S.N. (eds) *Governance in the extractive industries: Power, cultural politics and regulation.* London and New York: Routledge, 39–56.

Bos V. and Grieco K. (2018) L'eau: ressource naturelle, ressource politique? *Caravelle* 111: 59–78.

Bottaro L., Latta A. and Sola M. (2014) La politización del agua en los conflictos por la megaminería: Discursos y resistencias en Chile y Argentina. *European Review of Latin American and Caribbean Studies/Revista Europea de Estudios Latinoamericanos y del Caribe* 97: 97–115.

Budds J. (2010) Water rights, mining and indigenous groups in Chile's Atacama. In: Boelens R., Getches D. and Guevara Gil A. (eds) *Out of the mainstream: Water rights, politics and identity.* London: Earthscan, 197–211.

Camba A.A. (2016) Philippine mining capitalism: The changing terrains of struggle in the neoliberal mining regime. *Austrian Journal of South-East Asian Studies* 9 (1): 69–86.

Canterbury D.C. (2018) *Neoextractivism and capitalist development,* London and New York: Routledge.

Carrasco A. (2016) A biography of water in Atacama, Chile: Two indigenous community responses to the extractive encroachments of mining. *The Journal of Latin American and Caribbean Anthropology* 21 (1): 130–150.

Chin A. and Hari K. (2020) *Predicting the impacts of mining of deep sea polymetallic nodules in the Pacific Ocean: A review of scientific literature.* Deep Sea Mining Campaign and MiningWatch Canada.

Copeland N. (2019) Linking the defence of territory to food sovereignty: Peasant environmentalisms and extractive neoliberalism in Guatemala. *Journal of Agrarian Change* 19 (1): 21–40.

Coumans C. (2011a) Occupying spaces created by conflict: Anthropologists, development NGOs, responsible investment, and mining. *Current Anthropology* 52 (S3): S29–S43.

Coumans C. (2011b) Whose development? Mining, local resistance, and development agendas. In: Sagebien J. and Lindsay N.M. (eds) *Governance ecosystems.* London: Palgrave Macmillan, 114–132.

Cuvelier J. (2013) Conflict minerals in Eastern Democratic Republic of Congo. In: Hilhorst D. (ed.) *Disaster, conflict and society in crises: Everyday politics of crisis response.* London and New York: Routledge, 132–148.

De La Cadena M. (2010) Indigenous cosmopolitics in the Andes: Conceptual reflections beyond "politics". *Cultural Anthropology* 25 (2): 334–370.

de Theije M. and Bal E. (2010) Flexible migrants: Brazilian gold miners and their quest for human security in Surinam. In: Bal E., Eriksen T. and Salemink O. (eds) *A world of insecurity: Anthropological perspectives on human security.* London: Pluto Press, 66–85.

de Theije M. and Salman T. (2018) Conflicts in marginal locations: Small-scale gold mining in the Amazon. In: Lahiri-Dutt K. (ed.) *Between the plough and the pick: Informal, artisanal and small-scale mining in the contemporary world.* Canberra: ANU Press, 261–274.

Dolan C. and Rajak D. (2016) *The anthropology of corporate social responsibility,* New York and Oxford: Berghahn Books.

Eriksen T.H. (2018) Scales of environmental engagement in an industrial town: Global perspectives from Gladstone, Queensland. *Ethnos* 83 (3): 423–439.

Escobar A. (2008) *Territories of difference: Place, movements, life, redes,* Durham: Duke University Press.

Escobar A. and Alvarez S. (1992) *The making of social movements in Latin America. Identity, strategy and democracy,* Boulder: Westview Press.

Farthing L. and Fabricant N. (2019) Open veins revisited: The new extractivism in Latin America, part 2. *Latin American Perspectives,* 46 (2): 4–9.

Ferguson J. (2006) *Global shadows: Africa in the neoliberal world order,* Durham: Duke University Press.

Finn J.L. (1998) *Tracing the veins: Of copper, culture, and community from Butte to Chuquicamata.* Berkeley, Los Angeles and London: University of California Press.

Fraser A. and Larmer M. (2010) *Zambia, mining, and neoliberalism: Boom and bust on the globalized Copperbelt,* New York: Palgrave Macmillan.

Fultz K. (2016) *Economies of representation: Communication, conflict and mining in Guatemala.* PhD Dissertation, University of Michigan, Ann Arbor.

Garibay C., Boni A., Panico F., Urquijo P. and Klooster D. (2011) Unequal partners, unequal exchange: Goldcorp, the Mexican state, and campesino dispossession at the Peñasquito goldmine. *Journal of Latin American Geography* 10 (2): 153–176.

Gentilucci M. (2019) Self-determination referendum, mining and "interdependence": Reading the current political conjuncture in New Caledonia. *Small States & Territories* 2 (2): 157–170.

Graeter S. (2017) To revive an abundant life: Catholic science and neoextractivist politics in Peru's Mantaro Valley. *Cultural Anthropology* 32 (1): 117–148.

Grätz T. (2009) Colonial gold mining in Northern Benin: Forced labour and the politics of remembering the past. *Africana Studia* 12: 137–152.

Gudynas E. (2010) Diez tesis urgentes sobre el nuevo extractivismo. *Ecuador Debate* 79: 61–82.

Hausermann H., Amankwah R., Effah E., Ferring D., Mentz G., Atosona B., Chang A., Mansell C., Karikari G. and Sastri N. (2018) Land-grabbing, land-use transformation and social differentiation: Deconstructing "small-scale" in Ghana's recent gold rush. *World Development* 108: 103–114.

Hilson G. (2017) Shootings and burning excavators: Some rapid reflections on the Government of Ghana's handling of the informal Galamsey mining "menace". *Resources Policy* 54: 109–116.

Hilson G. (2019) Why is there a large-scale mining "bias" in sub-Saharan Africa? *Land Use Policy* 81: 852–861.

Hilson G. and Potter C. (2005) Structural adjustment and subsistence industry: Artisanal gold mining in Ghana. *Development and Change* 36 (1): 103–131.

Hilson G. and Yakovleva N. (2007) Strained relations: A critical analysis of the mining conflict in Prestea, Ghana. *Political Geography* 26 (1): 98–119.

Himley M. (2014) Monitoring the impacts of extraction: Science and participation in the governance of mining in Peru. *Environment and Planning A: Economy and Space* 46 (5): 1069–1087.

Hincapié S. (2017) Entre el extractivismo y la defensa de la democracia. Mecanismos de democracia directa en conflictos socioambientales de América Latina. *Recerca. Revista de pensament i anàlisi* 21: 37–62.

Horowitz L. (2002) Daily, immediate conflicts: An analysis of villagers' arguments about a multinational nickel mining project in New Caledonia. *Oceania* 73 (1): 35–55.

Horowitz L. (2010) "Twenty years is yesterday": Science, multinational mining, and the political ecology of trust in New Caledonia. *Geoforum* 41 (4): 617–626.

Huizenga D. (2019) Governing territory in conditions of legal pluralism: Living law and free, prior, and informed consent (FPIC) in Xolobeni, South Africa. *The Extractive Industries and Society* 6 (3): 711–721.

Jackson S.L. (2018) Abstracting water to extract minerals in Mongolia's South Gobi Province. *Water Alternatives* 11 (2): 21.

Kirsch S. (2014) *Mining capitalism: The relationship between corporations and their critics.* Oakland: University of California Press.

Kirsch S. (2018) Dilemas del perito experto: derechos indígenas a la tierra en Surinam y Guyana. *Desacatos* 57: 36–55.

Kumah C., Hilson G. and Quaicoe I. (2020) Poverty, adaptation and vulnerability: An assessment of women's work in Ghana's artisanal gold mining sector. *Area* 52 (2): 615–625.

Lahiri-Dutt K. (ed.) (2018) *Between the plough and the pick: Informal, artisanal and small-scale mining in the contemporary world*, Canberra: ANU Press.

Lassila M.M. (2018) Mapping mineral resources in a living land: Sami mining resistance in Ohcejohka, northern Finland. *Geoforum* 96: 1–9.

Lawrence R. and Moritz S. (2019) Mining industry perspectives on indigenous rights: Corporate complacency and political uncertainty. *The Extractive Industries and Society* 6 (1): 41–49.

Li F. (2015) *Unearthing conflict: Corporate mining, activism, and expertise in Peru*, Durham and London: Duke University Press.

Li F. (2016) In defense of water: Modern mining, grassroots movements, and corporate strategies in Peru. *The Journal of Latin American and Caribbean Anthropology* 21 (1): 109–129.

Linton J. (2010) *What is water? The history of a modern abstraction*, Vancouver and Toronto: UBC Press.

Long R., Renne E., Robins T., Wilson M., Pelig-Ba K., Rajaee M., Yee A., Koomson E., Sharp C., Lu J. and Basu N. (2013) Water values in a Ghanaian small-scale gold mining community. *Human Organization* 72 (3): 199–210.

Luning S. (2014) The future of artisanal miners from a large-scale perspective: From valued pathfinders to disposable illegals? *Futures* 62 (A): 67–74.

Luning S. and Pijpers R.J. (2017) Governing access to gold in Ghana: In-depth geopolitics on mining concessions. *Africa: The Journal of the International African Institute* 87 (4): 758–778.

Macintyre M. and Foale S. (2004) Politicized ecology: Local responses to mining in Papua New Guinea. *Oceania* 74 (3): 231–251.

Manky O. (2020) The end of mining labor struggles? The changing dynamics of labor in Latin America. *The Extractive Industries and Society* 7 (3): 1121–1127.

Marston A.J. (2017) Alloyed waterscapes: Mining and water at the nexus of corporate social responsibility, resource nationalism, and small-scale mining. *Wiley Interdisciplinary Reviews: Wires Water* 4 (1), e1175: 1–13.

McDonell E. (2015) The co-constitution of neoliberalism, extractive industries, and indigeneity: Anti-mining protests in Puno, Peru. *The Extractive Industries and Society* 2 (1): 112–123.

McNeish J.A. (2017) A vote to derail extraction: Popular consultation and resource sovereignty in Tolima, Colombia. *Third World Quarterly* 38 (5): 1128–1145.

Middeldorp N., Morales C. and van der Haar G. (2016) Social mobilisation and violence at the mining frontier: The case of Honduras. *The Extractive Industries and Society* 3 (4): 930–938.

Moody T.D. (with NdatsheV.) (1994) *Going for gold: Men, mines, and migration*, Berkeley, Los Angeles and London: University of California Press.

Moore J. and Velásquez T. (2013) Water for gold: Confronting state and corporate mining discourses in Azuay, Ecuador. In: Bebbington A. and Bury J. (eds) *Subterranean struggles: New dynamics of mining, oil, and gas in Latin America*. Austin: University of Texas Press, 119–148.

Nash J.C. (1979) *We eat the mines and the mines eat us: Dependency and exploitation in Bolivian tin mines*, New York: Columbia University Press.

Nelson D.M. (2013) "Yes to life = no to mining": Counting as biotechnology in life (ltd) Guatemala. *Scholar & Feminist Online* 11 (3).

Novo C.M., Bell S.E., Channa S.M., Pandey A.D. and Tuaza Castro L.A. (2018) Indigenous social movements in mountain regions. In: Kingsolver A. and Balasundaram S. (eds) *Global mountain regions: Conversations toward the future*. Bloomington: Indiana University Press, 83–118.

Ntewusu S.A. (2018) 10,000 miners, 10,000 votes: Politics and mining in Ghana. *Africa* 88 (4): 863–866.

Nuttall M. (2013) Zero-tolerance, uranium and Greenland's mining future. *The Polar Journal* 3 (2): 368–383.

Paredes Peñafiel A. (2020) *Desenhos da e com a natureza: os conflitos em torno dos projetos de mineração no Peru*. Rio Grande, Brazil: Editors da FURG.

Peet R. and Watts M. (2004) *Liberation ecologies: Environment, development, social movements*, New York and London: Routledge.

Perreault T. (2013) Dispossession by accumulation? Mining, water and the nature of enclosure on the Bolivian Altiplano. *Antipode* 45 (5): 1050–1069.

Pijpers R.J. (2014) Crops and carats: Exploring the interconnectedness of mining and agriculture in sub-Saharan Africa. *Futures* 62 (Part A): 32–39.

Prieto M. (2016) Practicing *costumbres* and the decommodification of nature: The Chilean water markets and the Atacameño people. *Geoforum* 77: 28–39.

Rasch E.D. (2012) Transformations in citizenship: Local resistance against mining projects in Huehuetenango (Guatemala). *Journal of Developing Societies* 28 (2): 59–84.

Rasch E.D. (2017) Citizens, criminalization and violence in natural resource conflicts in Latin America. *ERLACS* 103: 131–141.

Rasmussen M.B. (2015) *Andean waterways: Resource politics in Highland Peru*, Seattle and London: University of Washington Press.

Richards P. (2001) Are "forest" wars in Africa resource conflicts? The case of Sierra Leone. In: Peluso N.L. and Watts M. (eds) *Violent environments*. Ithaca and New York: Cornell University Press, 65–82.

Riofrancos T.N. (2017) Scaling democracy: Participation and resource extraction in Latin America. *Perspectives on Politics* 15 (3): 678.

Roder K. (2019) "Bulldozer politics", state-making and (neo-)extractive industries in Tanzania's gold mining sector. *The Extractive Industries and Society* 6 (2): 407–412.

Salas Carreño G. (2016) Mining and the living materiality of mountains in Andean societies. *Journal of Material Culture* 22 (2): 133–150.

Sawyer S. and Gomez E.T. (2012) *The politics of resource extraction: Indigenous peoples, multinational corporations, and the state*, London: Palgrave Macmillan.

Smith J.H. (2011) Tantalus in the digital age: Coltan ore, temporal dispossession, and movement in the Eastern Democratic Republic of the Congo. *American Ethnologist* 38 (1): 17–35.

Smith Rolston J. (2015) Turning protesters into monitors: Critical collaboration in the mining industry. *Society and Natural Resources* 28 (2): 165–179.

Sosa M., Boelens R. and Zwarteveen M. (2017) The influence of large mining: Restructuring water rights among rural communities in Apurimac, Peru. *Human Organization* 76 (3): 215–226.

Sosa M. and Zwarteveen M. (2012) Exploring the politics of water grabbing: The case of large mining operations in the Peruvian Andes. *Water Alternatives* 5 (2): 360–375.

Taussig M.T. (1980) *The devil and commodity fetishism in South America*, Chapel Hill: University of North Carolina Press.

Tschakert P. (2009) Digging deep for justice: A radical re-imagination of the artisanal gold mining sector in Ghana. *Antipode* 41 (4): 706–740.

Tschakert P. and Singha K. (2007) Contaminated identities: Mercury and marginalization in Ghana's artisanal mining sector. *Geoforum* 38 (6): 1304–1321.

Tubb D. (2020) *Shifting livelihoods: Gold mining and subsistence in the Chocó, Colombia*, Seattle: University of Washington Press.

Urán A. (2018) Small-scale gold mining: Opportunities and risks in post-conflict Colombia. In: Lahiri-Dutt K. (ed.) *Between the plough and the pick: Informal, artisanal and small-scale mining in the contemporary world*. Canberra: ANU Press, 275–293.

Urkidi L. (2011) The defence of community in the anti-mining movement of Guatemala: Defence of community in the anti-mining movement of Guatemala. *Journal of Agrarian Change*, 11 (4): 556–580.

Valladares C. and Boelens R. (2019) Mining for Mother Earth. Governmentalities, sacred waters and nature's rights in Ecuador. *Geoforum* 100: 68–79.

Velásquez T.A. (2012) The science of corporate social responsibility (CSR): Contamination and conflict in a mining project in the Southern Ecuadorian Andes. *Resources Policy* 37 (2): 233–240.

Velásquez T.A. (2018) Tracing the political life of Kimsacocha: Conflicts over water and mining in Ecuador's Southern Andes. *Latin American Perspectives* 45 (5): 154–169.

Velásquez T.A. (2019) From Canadian dragons to pristine *páramos*: Visual discourses in the campaign to defend Ecuadorian *páramos* from gold extraction. *Latin American and Latinx Visual Culture* 1 (4): 98–104.

Verbrugge B. and Geenen S. (2020) *Global gold production touching ground: Expansion, informalization, and technological innovation*, Springer Nature, Cham: Palgrave Macmillan.

Walter M. and Urkidi L. (2017) Community mining consultations in Latin America (2002–2012): The contested emergence of a hybrid institution for participation. *Geoforum* 84: 265–279.

Yagenova S.V. and Garcia R. (2009) Indigenous people's struggles against transnational mining companies in Guatemala: The Sipakapa people vs GoldCorp Mining Company. *Socialism and Democracy* 23 (3): 157–166.

Yankson P.W.K. and Gough K.V. (2019) Gold in Ghana: The effects of changes in large-scale mining on artisanal and small-scale mining (ASM). *The Extractive Industries and Society* 6 (1): 120–128.

Zhouri A. (2017) Introduction – Anthropology and knowledge production in a "minefield". *Vibrant: Virtual Brazilian Anthropology* 14 (2): e142072.

13

FINAL REFLECTIONS AND FUTURE AGENDAS

Robert Jan Pijpers and Dinah Rajak

This volume explores the anthropological contribution to understanding resource extraction. The chapters have traced the physical and technical landscapes of extraction, providing insight into how anthropologists examine the materialities, politics, and, above all, the social relations that drive and govern the exploration, development, and extraction of resources. Driving this enquiry are two key questions: what are the effects of resource extraction? And through what politics do they emerge? Responding to these, the contributors reveal the manifold ways that resource extraction is determined by and determines social relationships. The collection as a whole foregrounds how resource extraction is experienced; and how competing interests and power disparities shape those experiences of anticipation, negotiation, and contestation.

A central (though at times subterranean) seam running through the chapters are these questions of power, politics, and contestation generated by extractive processes – which represent major threads in the anthropological study of resource extraction (Appel et al. 2015; Gilberthorpe and Rajak 2016; Golub 2014; Kirsch 2014; Pijpers and Eriksen 2018; Rajak 2011; Rolston 2014; Shever 2012; Welker 2014). In this afterword, we reflect on the cross-cutting debates that emerge from the various chapters and offer some forward-looking thoughts, returning to that enduring fascination with *continuity and change* that continues to pattern anthropological interest in extraction. To bring the chapters into conversation with each other, we focus on three key interconnected themes: 1) heterogeneity, 2) connection/disconnection and inclusion/exclusion, and 3) conflict and resistance.

Heterogeneity: tracking diverse actors and processes across scales

In their annual review of the anthropology of mining, Ballard and Banks (2003) indicated that from the 1980s onwards a sense of a heterogeneous mining

DOI: 10.4324/9781003018018-13

community developed. This concerned both the incorporation of different actors (such as NGOs, legal agencies, and media) into the "triad stakeholder model" of state, corporation, and community, as well as a growing awareness of the internal complexities of specific actors. The chapters of this volume illustrate that expanding and evolving constellation of actors that constitute extractive processes, thereby breaking open what are so frequently represented as transnational structures fuelled by an agentless logic of capital accumulation and insatiable scramble for resources, to reveal the complex agencies at work within them. They draw attention to the diversity of actors (corporate, collective, and individual) and agencies (human and non-human) that inhabit and shape extractive environments.

In their chapter on Water and Conflict, Li and Velásquez, for example, highlight the heterogeneous yet globally connected character of coalitions and alliances of resistance to extraction. In doing so, they foreground internal ambivalences and contradictions. How, for instance, people's dependence on water does not preclude aspirations of employment at the mine whose operation diverts and imperils the vital source of (clean) water. In this sense, people may simultaneously oppose extraction and desire the jobs it offers. Shifting the anthropological lens back to the corporate domain, Shever's chapter explores the internal complexities of corporations, arguing that the corporate effect "generates businesses that are autonomous, coherent, and bounded individuals in some respects, yet diffuse, heterogeneous, and inconstant assemblages in others" (page 33). Observing these practices of corporate and individual shape-shifting and identity-making, raises the question what is being performed, to which public, and in whose interests?

The heterogeneity that characterizes relations between actors within extractive communities, is reflected in the systems and practices that structure resource extraction. As Cuvelier, Geenen and Verbrugge underscore, extractive governance is marked by "ongoing process of hybridization, whereby distinctions between formal and informal, modern and traditional, state and non-state, and local, national, and transnational governance are becoming increasingly difficult to maintain" (page 87). These processes also have complex vertical dimensions. Indeed, as Luning's chapter illustrates, the practices of governing surface and sub-surface space may not be in sync. Several authors draw attention to the equally multi-faceted, attenuated, and at times disjointed and dislocated, dynamics of the operating systems and practices at work in resource development. Shever, for instance, observes a corporate shift from producing resources to managing complex infrastructures of subsidiaries, contractors, and subcontractors who do the actual work. The relegation and evaporation of clear responsibilities and the increasing number of short-term and precarious contracts are just two examples of the many effects of these processes.

Moving from organograms to the coal face, several authors explore how formal and informal hierarchies of power are inscribed at extractive sites. Illustrative in this regard is D'Angelo's discussion, which highlights the gendered and racialized divisions of labour at the workplace and how these are embedded in wider social stratifications and political structures such as the labour repressive economy of apartheid South Africa. Such observations stress how, at multiple levels, systems and practices are subject to and shaped by local conditions and national politics as much as by the codes and protocols of global governance and corporate strategy as often assumed in transnational commodity chains.

Understanding dynamics of connection/disconnection and inclusion/exclusion

The chapters highlight the power of ethnography not only to capture the intricacies of these heterogeneous and highly politicized fields of extraction, but the ways that actors connect with (and are alienated from) resources – whether material, political, or symbolic – within them. Through processes of extraction, commodification, and capitalization, competing systems of value are created and eroded across different contexts. The discussions of Ferry and Luning are telling in this regard. In her chapter on Materiality and Substances, Ferry recounts how silver in Mexico is seen to leave its place of origin as an article of trade, but at the same time remains in its place of origin in "the form of miners' houses, city plazas, and the jobs" (page 106). Along similar lines, Luning analyses relations between people, resources, and their environment as Polanyian processes of embedding and disembedding, foregrounding that "making chunks of earth into commodified resources depends on a double movement of disembedding mining matter from the wider landscape by embedding it in social circumstances" (page 189). Taking a different perspective, D'Angelo highlights the role of mobility and the establishment of (voluntary and involuntary) connections between different places and people with different expertise in the diffusion, innovation, and development of mining technologies. Finally, as Bainton and Skrzypek emphasize, processes of connection and disconnection are of crucial methodological importance too. Ethnographer's engagements and disengagements with fields (and cycles) of extraction and the interlocutors they work with, whether investigating the financial flows that fuel extraction or the physical work of pulling rock from the ground, can equally be seen as a series of connections made, broken, and re-made, often over a wide geographical and temporal span (see also, for example, Gilbert 2020).

Questions of connection and disconnection are of course profoundly political as they drive, disrupt, or reinforce the processes of inclusion and exclusion which invariably characterize extractive investments (Ferguson 2005; Gardner 2012; Kirsch 2014; Rajak 2011). Several of the chapters consider how these exclusionary processes are spatialized, carving "resource frontiers" into minescapes built on new conurbations which incorporate some while dispossessing others of land and livelihoods, or leaving them on the margins of new resource-centred developments, and the opportunities they may bring. As Nuttall shows, the ascription of places as "remote and pristine, empty and wild, and how they become subject to techniques and practices of conquest, commodification and export-oriented resource extraction" (page 60). These classifications prepare the ground for investment/incursion and development/extraction, and the inevitable division of spoils which privileges some interest groups (those amenable to the extractive agenda), while sidelining or silencing others whose positions may be less acquiescent. All point to the discursive and material that goes into establishing resource enclaves in which physical or symbolic walls define boundaries between what is and who are inside and what is and who are outside the extractive space (discussed in Shever's chapter). From geo-prospecting to resource development to withdrawal, impact assessments are increasingly key tools used by corporations to mark out the territory and make legible the locality. As Lanzano discusses in his chapter, these too work as instruments of exclusion, rendering certain data, phenomena, and people more visible than others.

Crucially, corporate tools of the trade such as social and environmental impact assessment (see also Lanzano's chapter) or stakeholder analysis direct and distribute resource flows and revenues, determining who can (and who can't) get on the end of the (often scant) rewards that trickle out from extractive operations. Beneath the technical mechanisms of stakeholder engagement and participatory development invoked by extractive companies as the near sacred canon of corporate social investment, access to the revenues and opportunities provided by resource development are patterned by entrenched hierarchies of status, and questions of ownership, belonging, autochthony, gender, and inequalities of education and social capital constrain processes of redistribution, a debate which is central to Pijpers' chapter.

Conflict and resistance

This brings us to questions of conflict and resistance – the third central theme cross-cutting the collection and a rich seam in anthropology's long engagement with resource extraction. More specifically, as the various chapters here highlight, a key contribution of anthropological perspectives (facilitated by grounded ethnographic approaches) has been to reveal the often hidden intricacies of contestations over resource extraction and how these play out between local, national, and global arenas. Whether localized conflicts between community and company over water access and pollution, as discussed in Li and Velásquez' chapter, or struggles over authority and agency by indigenous groups and artisanal miners contesting their dispossession of land and livelihoods by powerful corporate–state coalitions (Nuttall and Cuvelier et al.'s chapters respectively), conflict has been a vital concern in the anthropology of resource extraction. That commitment to confronting head on the politics and power relations at play, to identifying what's at stake, and for whom, and giving voice to groups alienated through mechanisms of appropriation, privatization, and formalization has been at the forefront of anthropologists studying resource extraction as it is in this volume.

Omnipresent in all of these discussions, is the key role of power and the patterns of access and control that derive from it. How is power claimed, legitimized, established, exercised, challenged, reproduced, and reconfigured? What forms of access and control do they enable or constrain? And crucially, what are the evolving weapons of the weak (cf. Scott 1987) marshalled in attempts to resist practices of domination and dispossession exercised in the name of resource development? In that light, Rubbers draws attention to a body of work that unearths mineworkers' everyday forms of resistance targeted at their corporate employers – forms of sabotage, slowdowns, and strikes – which seek at times overtly at others, covertly, to contest the adverse conditions and everyday violence of their employment. Meanwhile Pijpers' chapter explores a more explicit enact-ment of violence as young Papua New Guinean men challenge the landed power relations in a bid to access extraction-related benefits, denied to them in formal politics of distribution.

Future agendas – continuity and change

The new millennial scramble for resources promised to rehabilitate extractive capitalism and banish the resource curse with transformations in global governance and local content agreements (see for example Collier 2008; Melber and Southall 2008). Yet as this collection shows, anthropologists continue to chronicle (among a wide range of phenomena and dynamics) experiences of dispossession, disen-franchisement, domination, and exclusion for those living in the wake (or antici-pation) of extraction. This leads us to conclude by asking, is this the same story, the same cycle, of boom and bust, hope and betrayal on repeat, or is there something specific about how the resource curse – its opportunities and its costs – plays out in relation to millennial capitalism?

The enduring phrase of Max Gluckman and the Rhodes-Livingstone anthro-pologists resonates strongly: continuity and change. Taking this temporal frame-work as a heuristic device, offers an important theoretical and historical overview to processes of investment and resource extraction today. It is neither the dawn of a bold new era of business-led development for the continent superseding state impotence and failures of the aid industry, nor merely "old wine in new bottles" (Melber 2007 quoted in Melber and Southall 2008: 12), as new entrants on the extractive landscape reinvigorate century-old forms of imperial domination, com-pradore co-option, and ultimately the dispossession of the poorest. At the same time, the current orthodoxies of good governance, inclusive growth, and sustain-able extractives should not be seen merely as smokescreens for the imperial endeavours and mercenary pursuit of resources, but rather as part of the moral apparatus that sustains and enables the scramble for resources.

Subsequently, as we confront the escalating climate emergency (albeit too slow and too late) the imperative to transition to a low-carbon economy will require radical transformations in the extractive industries and financial markets. As Bainton and Skrzypek discuss in detail, the role of anthropologists, studying, challenging, or in some cases working with processes of resource extraction, will also need to adapt. The various authors in this volume have all reflected on this evolving role, putting forward suggestions for future directions for the anthropology of resource extraction. These comprise both a continuation of existing lines of inquiry, but also

their expansion by incorporating new questions, new forms and spaces of extraction, and innovative (methodological) approaches with which to investigate them.

Several of the authors in this volume have raised questions that would expand
the field of analysis within the specific chapter themes. D'Angelo for example asks
to what extent and how processes of automation, including artificial intelligence,
will influence both large and small-scale mining in the future. Will robots, green
technologies, and biotechnologies ever solve (some of) the problems surrounding
resource extraction, particularly in terms of social and environmental impact? Or
will they simply contribute to intensify and expand mining frontiers pushing (or
hiding) these same problems elsewhere, in the depth of the oceans or, maybe, on
the lunar and asteroid surfaces? Likewise, Rubbers (this volume) urges ethnographers of mining to pay more attention to the interactions between mineworkers
and machines as the latter take on/over ever more of the tasks of the former, with
major implications for these labour-intensive industries and the jobs they provide.
Crucially, Rubbers stresses, labour policies and their broader social dynamics should
feature more prominently in anthropology, as the terms and conditions in which
mineworkers are employed become ever more uncertain; a concern connected to
Cuvelier, Geenen, and Verbrugge's observation of the need for more attention to the
governance of labour markets, as part and parcel of a more robust anthropological
engagement with the various state bodies that are involved in governing extraction,
but to a large extent remain an unopened black box in ethnographies of extraction.

Other contributors focus less on change (and continuities) *within* the organizing
systems and apparatus of extractive industries, and more on capturing how resource
worlds evolve in relation to external pressure from new global crises, whether the
covid-19 pandemic, or shifting national politics and ideologies (see Pijpers this
volume). Recent examples concern extractive practices fuelled by the pursuit of an
energy transition and green economy agendas, such as lithium extraction for the
production of batteries or the mining of copper; or the effects of pro and anti-
mining politics, such as the pro-coal position of former US President Donald
Trump or the anti-mining position that led to the 2021 election victory of the
Inuit Ataqatigiit (IA) party in Greenland. Moving from the local specificity that is
the anthropologist's usual stock and trade, Nuttall suggests a need to scale up, and
pay greater attention to global interconnectedness, resonating with Arboleda's
(2020) idea of "the planetary mine", which foregrounds the planetary scale of the
movements and infrastructures of extractive operations. To this debate of the planetary mine, Luning argues, "anthropologists can contribute substantially with their
patient and meticulous attention to how infrastructures are made and used in the
daily, the routine, and the repetition of social practices" (page 198).

As many contributors highlight, innovations in methodological approaches or
positions are and will be a key concern for the anthropology of resource extraction
in the years to come. This demands paying more attention to the complete scope
of senses – sight but also touch and hearing – to achieve a genuinely miner-centred
perspective on the experience and lifeworlds of mineworkers, as Luning stresses. Or
as Ferry emphasizes, a sustained focus on the temporalities of extraction to draw the

pasts, presents, and futures of extraction into a coherent frame. Lanzano also stresses the importance of time, specifically arguing that notions of sustainability must be de-centered; an agenda which will benefit from a further expansion of the work of several anthropologists on the temporalities and local practices of future-making (see also D'Angelo and Pijpers 2018; Ferry and Limbert 2008). Meanwhile, Nuttall stresses the necessity of bringing anthropology into conversation with other disciplines, considering how extraction has left and still leaves lasting traces on Earth processes and geologies. Finally, Shever's analysis of corporations as being made and understanding them "as potent and pervasive, but not as invincible or indestructible" (page 34), enables her to argue that corporations can therefore also be made differently. In doing so, Shever assigns a clear task to anthropologists studying global extractive capitalism. Whether one agrees with this particular viewpoint or not, there is no doubt that the position of anthropologists, and the discipline as a whole, are crucial parts of scholarly engagements with the contested terrains of extraction, as Bainton and Skrzypek observe, and reflecting on these positions, they reassure us, should be a more structural element of anthropological writing.

In recent years, anthropologists eager to confront and capture the fast evolving terrain of extractive capitalism, and the socio-political and ecological crises it produces, have pushed the boundaries of ethnographic inquiry into new spaces in so-called global corridors of power following the digital worlds of resource finance (Ferry this volume); fossil-fuel divestment campaigns (see for example Corder 2021); shareholder activism in global oil corporations demanding greater climate accountability (see for example Rajak 2020); world bank inspection panels; and international arbitrations in disputes between sovereign nations and mining multinationals (see for example Gilbert 2018). Such spaces present new problems of access requiring new methodological tools to gain entry (Müftüoglu et al. 2018). At the other end of the spectrum, we find ethnographers pushing to penetrate new spaces of extraction, as extractors pursue frontiers to satisfy investor and consumer appetites for more and novel resources. Spaces which present equal problems of access: deep waters and ocean floors (see D'Angelo, Li and Velásquez, and Luning this volume); new layers of the underground (see Luning this volume); and of course spaces associated with urban mining and processes of recycling, as well as the environments that become accessible with the melting of Arctic ice caps. These new frontiers break new ground not only geologically and technologically, but also politically, generating new struggles, contestations, and fronts in anti-extraction protest. In the coming years, a critical challenge for the anthropology of resource extraction will be to figure out our role (and responsibility) in relation to these movements, intellectually, politically, and ethically.

Bibliography

Appel H., Mason A. and Watts M. (2015) *Subterranean estates: Life worlds of oil and gas*, Ithaca: Cornell University Press.

Arboleda M. (2020) *Planetary mine: Territories of extraction under late capitalism*, London: Verso Books.

Ballard C. and Banks G. (2003) Resource wars: The anthropology of mining. *Annual Review of Anthropology* 32: 297–313.

Collier P. (2008) The profits of boom. Will Africa manage them differently this time? *Optima* 54 (December): 2–7.

Corder M. (2021) Dutch court orders Shell Nigeria to compensate farmers. The Associated Press, 29 January. Accessible through: https://apnews.com/article/business-netherlands-nigeria-the-hague-pollution-df365847d4cf6bf2a1fcd1b94d1cbf2e.

D'Angelo L. and Pijpers R.J. (2018) Mining temporalities: An overview. *The Extractive Industries and Society* 5 (2): 215–222.

Ferguson J. (2005) Seeing like an oil company: Space, security and global capital in neo-liberal Africa. *American Anthropologist* 107 (3): 377–382.

Ferry E.E. and Limbert M.E. (2008) *Timely assets*, Santa Fe: School for Advanced Research Press.

Gardner K. (2012) *Discordant development: Global capitalism and the struggle for connection in Bangladesh*, London: Pluto Press.

Gilbert P. (2018) Sovereignty and tragedy in contemporary critiques of investor state dispute settlement. *London Review of International Law* 6 (2): 211–231.

Gilbert P. (2020) Speculating on sovereignty: "Money mining" and corporate foreign policy at the extractive frontier. *Economy and Society* 49 (1): 16–44.

Gilberthorpe E. and Rajak D. (2016) The anthropology of extraction: Critical perspectives on the resource curse. *The Journal of Development Studies* 53 (2): 186–204.

Golub A. (2014) *Leviathans at the gold mine: Creating indigenous and corporate actors in Papua New Guinea*, Durham, NC: Duke University Press.

Kirsch S. (2014) *Mining capitalism. The relationship between corporations and their critics*, Berkeley: Stanford University Press.

Melber H. and Southall R. (2008) *The new scramble for Africa*, Scottsville, South Africa: KwaZulu Natal Press.

Müftüoglu I.B., Knudsen S., Ragnhild F.D., Eiken O., Rajak D. and Lange S. (2018) Rethinking access: Key methodological challenges in studying energy companies. *Energy Research & Social Science* 45: 250–257.

Pijpers R. and Eriksen T. (2018) *Mining encounters: Extractive industries in an overheated world*, London: Pluto Press.

Rajak D. (2011) *In good company. An anatomy of corporate social responsibility*. Palo Alto, CA: Stanford University Press.

Rajak D. (2020) Waiting for a *Deus ex Machina*: Sustainable extractives in a 2 degree world. *Critique of Anthropology* 40 (4): 471–489.

Rolston, J. (2014) *Mining coal and undermining gender: Rhythms of work and family in the American West*, New Brunswick, NJ: Rutgers University Press.

Scott J. (1987) *Weapons of the weak: Everyday forms of peasant resistance*, New Haven and London: Yale University Press.

Shever E. (2012). *Resources for reform: Oil and neoliberalism in Argentina*, Berkeley: University of California Press.

Welker M. (2014) *Enacting the corporation. An American mining firm in post-authoritarian Indonesia*, Berkeley: University of California Press.

INDEX

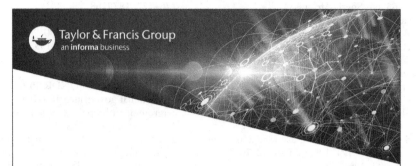

Printed in the United States
by Baker & Taylor Publisher Services